THE FUTURE OF PROPERTY

HOW TO LEVERAGE EMERGING TRENDS & TECHNOLOGY TO TAKE YOUR PROPERTY BUSINESS TO THE NEXT LEVEL

KYRON GOSSE

The Future of Property

First Printing: 2018

Best Seller Book Design

Book Design by Adele Wiejaczka

www.bestsellerbookdesign.com

KYRON GOSSE

www.thefreedominvestor.com

ORDERING INFORMATION

Special discounts are available on quantity purchases by corporations, associations, educators, and others. For details, contact the publisher at the above listed address.

BOOKING, PRESS & SPEAKING INQUIRIES

www.thefreedominvestor.com/contact

DISCLAIMER

The information contained in this book and any form of communication related to this book is for information purposes only. The purpose of this information is for educational purposes only and is not specific information for your personal situation. Neither Kyron Gosse nor 'The Freedom Investor' network of companies hold themselves as providing financial, legal or any other type of advice. Kyron also does not make any recommendations or endorsements to any investment, adviser or other service or product (including those used as case studies inside this book). In addition, Kyron does not offer any advice regarding the nature, potential value, or suitability of any investment, security or investment strategy. The material in this book does not constitute as advice and you should not rely on any material in this book to make (or refrain from making) any decision or take (or refrain from taking) any action. All investment, borrowing or tax decisions should be made with the advice of an appropriately licensed or qualified adviser.

TABLE OF CONTENTS

Extra

I have also created an online course that works hand in hand with this book. If you are a person who learns better watching videos, or from a more hands on medium, then you might want to consider following along online.

www.freedom-investor.thinkific.com/courses/future-of-property

Prenote

Welcome to the Future of Property. I am really excited that you are going to be joining me on this journey, as we take a look at how future trends and technologies are going to affect, and ultimately disrupt, the real estate industry.

This book is perfect for existing property investors, real estate professionals and property entrepreneurs who are looking to add an extra edge to their business. It is for those who are looking to find out ways that they can add an aspect to their business, to accelerate it and take them up to the next level.

If you're a new investor, this is also going to help you to start with fresh ideas and see how things are happening in today's market, so that you do not have to go down the old slow path. You can just start new, start fresh and accelerate to get returns that have taken other investors years to achieve.

I wrote this book to give you awareness of what the future five waves are going to be and how they are going to be affecting the real estate industry in different ways. It is going to allow you to level up to implement team, trends and technology into your business so that you can operate on a whole new plane. You are going to learn how you can spot these emerging trends and technologies, as well as what other waves are coming in the future along with mini-waves within the waves. As a result,you will be able to understand and start to look out there by yourself to see what is happening and take advantage of these waves. Ultimately, you will get the 'unfair' advantage over all the other property investors out there.

You are also going to learn how you can future-proof your business, so that no matter what happens within the market and industry, you are protected.

Then finally, you are going to be able to implement a new exponential business model utilising the 3T system: team, trends and technology.

This book has been two years in the making as I had to learn the hard way how to grow my business from being the one running around like a chook with no head, to managing an empowered team of entrepreneurs who are all taking charge of one aspect of the business and moving it forward. As with most entrepreneurial journeys, I have had my peaks and my troughs, my wins and my losses and I have to say: nothing had prepared me for the difference between going from a simple property investor to running a property business – especially a global one at that.

Through blood, sweat and tears (as well as the odd temper tantrum) I have had to leverage my mentor's wisdom, seek answers in the strangest of places, and juggle the reality of being an international investor and business person whilst trying desperately to keep a firm grip on my identity and mantra as The Freedom Investor – that is maintaining my freedom whilst growing three separate companies.

To everyone who has played a part in helping me to get to this point, I want to say thank you. To those people who have endured my crazy ideas and more importantly those who have helped me to turn those ideas into reality, I say thank you. And to those who have put your trust and faith in me, by investing your hard earned cash into my ventures, again I say thank you.

And to you the reader, I want to thank you for taking the time to read this book and educate yourself on new ideas, on having an open mind to look at property from a different point of view, and to be doing something to help yourself be better off financially and working to create the life you want to live. After all, in the words of Howard Thurman: "The world needs more people who have come alive".

INTRODUCTION

It has been nearly three years since I sold all my possessions, bought a plane ticket and moved to the other side of the world without any idea of where I was going to live, or where my journey was going to take me. All I knew was one thing: I was finally free and property had gotten me there.

In my first book, The Freedom Investor Blueprint, I laid out the plans on how I had created my freedom and how other investors looking to create more freedom in their lives could leverage the same system to be able to live their dream life.

Fast forward two years and having helped thousands of investors with their freedom portfolios, I find myself at a crossroad – as I do not actually own any properties at the moment.

I sold up my properties at the peak of the market when I saw the trends were leading to a downturn in the market and I wanted my cash to work harder than I did. Therefore I was able to sell whilst there was still a lot of demand for property and then I found ways to invest my money through partnerships and joint ventures, being rewarded every few months with a large cash injection. Because of this I had the freedom to travel the world.

In that time, I have rode camels through the Sahara, fallen in love with Agua De Valencia, rode a motorbike through Northern Vietnam, drank my way across Siberia in a train with a group of Digital Nomads, been caught in a snowstorm whilst riding in Mongolia, attended a conference of Futurists in Beverly Hills, drank tea with the cave-dwelling Berber people in Morocco, helped Syrian refugees in Belgrade, drank Port in Porto, practiced yoga in Bulgaria, spotted Cheetah, Lions and Leopards in the Masai Mara, Elephants in the Okavango Delta and hippos in Lake Victoria, all before finally settling down in Bali (for now). Let's just say, life has been pretty good.

Unfortunately though (or maybe I should say fortunately), I get bored easily. Whilst all these amazing travels were happening, I was still working on my businesses – with the aim of helping other investors create freedom and build wealth through property. I did this through education and through co-investing with them. I faced many challenges trying to run an international business from my laptop. Many a time I had to reschedule a call because I was going to be flying, and I've even scheduled an 8-hour layover in Hong Kong just to be able to present a webinar) On more than one occasion I also woke up at 3 am to attend a conference call. The thing is, even though I had been successful in the property industry and created my dream life, I was still self-employed. I was still the person who had to pedal the bike if I wanted to actually get anywhere.

Sure, I could have given up and solely focused on my investments, living off the profit from them. But the thing is, I feel alive when I see other investors achieving their goals and living a life of awesome.

I remember working with one particular client, Nicholas, who had made it his mission to donate half of everything he made from working with me to worthy causes (this was on top of the missions he was already doing in Africa). In the first year, he donated close to $50,000 to different charity groups and families in his community who were in need. I remember seeing how empowered he was, how in flow he was and most importantly how he simply came alive. I knew there and then that through helping people like Nicholas, I could make my own impact on the world and so I vowed to use my skills to help as many people create freedom and build wealth as possible so that they too could gain financial stability and solve some of the problems in the areas they are passionate about. Whether it was simply donating money, time and skills in areas that were in need (like Nicholas did) or whether it was to take on a much bigger challenge and solve a global issue, be part of a social movement.

I used this motivating factor to build my companies from the ground up, with a strong purpose and culture based around a mutual belief that everyone on this planet is born to do something great. The problem is, many of us get caught up in the rat race and trying to survive. My goal is through my companies to help the whole world come alive and follow their dreams.

I had to find a path to achieving this goal, and I already had a massive love of property, so I thought it was probably the most appropriate way to proceed. Upon discovering many of the factors I will be covering in this book, I decided I would help people to create their financial stability so they could come alive, by working to democratise real estate and make it accessible to all, so that everyone in the world can leverage real estate as their wealth creation vehicle of choice.

Across my companies we offer the tools and resources one needs to create their financial stability and quit the daily grind; from financial education, to investment opportunities, better funding strategies and even co-owning properties around the world, everything I do in my businesses are to help make property more accessible for the common man.

To take it one step further, I am even making it my own personal mission to see that the global homeless problem is solved and ensure that everyone in the world has a roof over their head. Doing so provides a stable framework for people to then go follow their dreams and follow their bigger purpose. Imagine knowing that if you wanted to pursue your passions and dreams you had a safety net under you, ensuring that if it all failed, you wouldn't have to live out of your car, that your family would always have food on the table, and that electricity would always be available.

But first, I needed to find a better way of doing business other than simply being self-employed. I knew if I could leverage myself better, I could reach more people and make more of an impact. This is where I believe the power of entrepreneurship comes in. In the Freedom Investor Blueprint, I point out the benefits of tying together traditional property investment and entrepreneurship to become a property entrepreneur and leverage entrepreneurship principles to create your freedom and build your wealth in the property industry. The more I ventured down the entrepreneur track, the more I spotted bigger and better opportunities appearing in the industry. I watched Airbnb become the world's largest real estate company – and it owned no physical property at all.

It was there and then I realised there was a whole wave of trends and technologies emerging that were going to seriously impact the real estate industry and if we could see them, if we could prepare for them and if we could ride them, then there was not only a tonne of money to be made, but we could even steer the direction of the waves to do some serious awesomeness in the world. So, I decided to focus my attention on helping people just like Nicholas, who had created success in property investing and were now working within their property business and dreamed of making bigger impacts, doing more good and building their business to be even greater.

It is easier than ever to start small and make impacts on causes you believe in. In fact, in order to thank you just for reading this book and even thinking about how you can make positive impacts on the world, I am going to leverage the **www.b1g1.com** platform and send 10 bricks to a family in need in India to help them get a new lease on life and build their first property. See just how easy it is to make a difference and do awesome?

Learn More:

If you like what you are reading, there are a few places you can go to learn more.

The first place you should head to is the Future of Property website, **www.futureofproperty.co**. Here, you are going to find lots of articles and timely updates on what is happening in the market – including some of the trends and technologies that are emerging – so that you can keep in the loop and know what is happening across the world with future trends and technologies that are influencing the property industry.

The second thing is to connect in with other property entrepreneurs and other futurists, to make sure you know what's going on in the different technologies around the world, find out what is happening in your area, and connect with other investors. Make sure you join the **Future of Property Facebook group**.

Another place you can go, if you want to learn more about me, is **www.thefreedominvestor.com**. Of course, I am obviously across all the social media platforms, from Facebook to Twitter to YouTube and LinkedIn. If you want to learn more, go check them out and make sure that you connect up with me. Send me a message, say, "Hey Kyron, I'm reading your book right now." I always love to meet people in my community who are doing awesome things.

A Note on How to Read This Book:

As you read this book, you are going to find that a lot of the content in here is very contextual. There is a lot of things that are far into the future that may not affect you for a long time, and depending on how technology and trends play out, some of it might not come into being, and some of it might affect us in different ways. It is impossible to tell how it is going to affect us overall.

Some of what you read is going to be just nice to know. It is nice to know what will happen in the future, what trends will come this way, and what technologies are being created so you can keep it in the back of your mind for when things do start appearing. Ultimately, however, you are the one that needs to make the decision of how you implement all the content in this book.

There are two ways that you can go about this. The first way is keeping with your existing business and then looking for the trends that are going to help you accelerate your business, the technologies that are going to help you leverage your existing business and how you are helping your customers and being of service to them in a much better way.

The second way you could do this is to find a trend or to find a technology that you like, jump on board, and build a whole new business around that technology or around that trend. As we go forward, keep that in the back of your mind and just decide which way you want to go. Are you using this information to supplement an existing business, to pivot an existing business, or are you looking for something that is going to allow you to launch a whole new business around that technology?

CHAPTER ONE

VISITING THE FUTURE

A PROPERTY ENTREPRENEUR'S GUIDE TO TIME TRAVEL

Time travel has long been a thing of science fiction, with Hollywood playing their part with all sorts of plot twists and unexpected endings in their movies. However, when it comes to time travel in reality, the ending is not so unexpected – in fact, the whole point is that it is expected, hence why we time travel in the first place: because it gives us the ability to know what to expect and predict the future.

Okay, so I am obviously not talking about time travel as in machines with lights and sirens jumping across the timesphere, but it is possible to go look at different markets around the world, especially ones that are more advanced or younger than your own, in order to adapt some of the strategies that are working there into your own market.

For example, Australia and New Zealand have a much younger property market than the United States or England. Therefore, if we look at what is working in England, such as commercial to residential conversions, HMOs or multiple people living in one house, these sorts of things really are not that common in the Australian or New Zealand market at the moment. This gives you the ability to adapt that strategy for the NZ market and gain the first mover advantage.

Or, we could check out America where you can get a lot of vendor finance and it is easier to do creative deals. Again, that sort of thing is not that commonplace in Australia or New Zealand. It is possible to look at what is working around the world and say, "Hey, could that be adapted to my own market?," and basically go from there.

Now, the way you do this is to look at a more advanced market than your own, look at a bigger market than your own, or look at someone who is facing a bigger problem than your own. For example, again, if we want to look at some of the problems that maybe London is facing, where there is a lot of people trying to live in a small city, we can look at countries like Hong Kong or Singapore who have a much higher population density, and see what are some of the issues that they have faced to come up with a way to house millions of people on this tiny little island. These solutions can then be adapted to the London market, and the problem is solved with the entrepreneur making a great deal of profit.

Bangladesh is another perfect example. Bangladesh has many rivers. It also has big problems with flooding. But now they are working with the Dutch government, (26% is below sea level, and 50% is only 1 metre above sea level) to implement many of the solutions that have worked in The Netherlands into Bangladesh so that they can solve these issues. Just by thinking like that, 'who has a bigger problem and how are they solving it?'

Perfect example: Beijing working alongside Singapore. Singapore is the best example right now of a highly advanced living city, which has good work-life balance and has some sort of focus on the environment. Whereas Beijing, obviously, it is known for having a lot of pollution. They are trying to clean that up, so they are mimicking a lot of what Singapore is doing in Beijing. They are looking to clean up all the smog, looking to clean up the pollution and ultimately, as you will find out later, trying to become a living green city.

One of the best examples of adapting something to your market that works overseas is crowdfunding. As you can see in the graph, crowdfunding really launched in the US and has been particularly effective in Europe, particularly the UK market. Look at Australasia on the graph, and it shows that crowdfunding has not yet taken off. Whilst these stats are a little bit old – they are from 2015 – it still shows there is a lot of potential to take what has worked in the US and Europe and simply introduce it into the Australasian market. This would allow for fast growth and the potential to monopolise the market. In the same way, we can see on the Asian bar of the graph, there has been massive growth in recent years, this is a solid, well-established trend that an entrepreneur could immediately jump on, simply copying the existing business models from Europe or America.

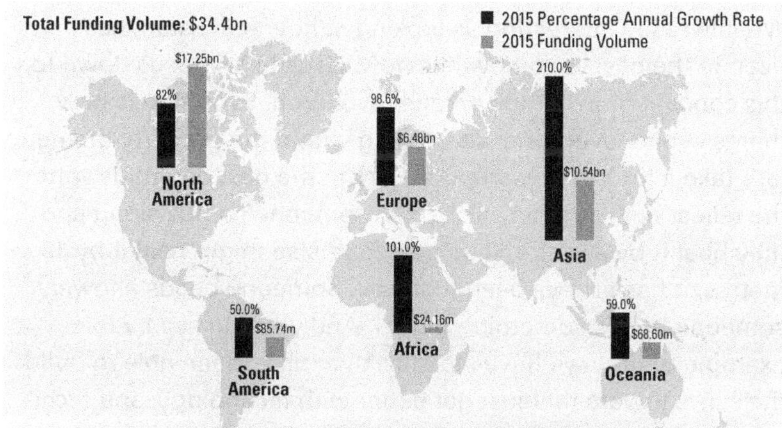

GROWTH BY CROWDFUNDING REGION PREDICTION
FOR 2015 IN MILLIONS OF USD (RESEARCH BASED ESTIMATE)

When coming from a younger market, you can simply copy what other people are doing and just do it better or do it suited to your own market. You might have heard of 'We Work', a massive coworking company based out of the United States. It has a multi-billion dollar valuation range. A company in China was not even that original when they ripped it off. They called it 'UR

Work', and now it is also a multi-billion dollar company. Simply by taking a business model and a trend that was working in the USA, adapting it to the local Chinese market, and just running with it. There is so much that is possible.

If you are coming from a more mature market, you do not really have the ability to run around saying: "Hey, look, I am going to adapt that into my market." But you can take what is working in your market to a younger market where people are yet to adopt the strategy, trend or technology. Another option, if you are in a more mature market such as the USA or England, where you are already leading the way, you should be focusing more on the technology side, so that you can understand how you can leverage what you are already doing or what is already working in your country with technology.

What is really interesting, is looking at how real estate has been in the past as to how it is now. It really brings us down to this concept of evolution versus revolution. Evolution is how things evolve over time, so keeping within the property theme, let's take a look at skyscrapers. Back in the day, pyramids were the tallest buildings on earth, then someone comes along and they beat it by a metre, then someone else might beat it by 10 metres, it has always been that way. Someone builds a tower, someone builds something bigger. And you can see by this example of how we have steadily, over time, been able to build the same towers that just get better with technology, and architects pushing the limits on how can they can just get that extra 10 metres. How can they just keep building so that they are a little bit taller than all the other buildings around the world. You will notice, if you have a look around the world, the tallest building title is generally only held for 10 or 20 years until someone comes along and says: "We are going to beat that."

Revolution, on the other hand, is starting from scratch, and instead of saying: "How can we get that extra 10%?" It is saying, "How can we 10 times it?" The best example I was given is by

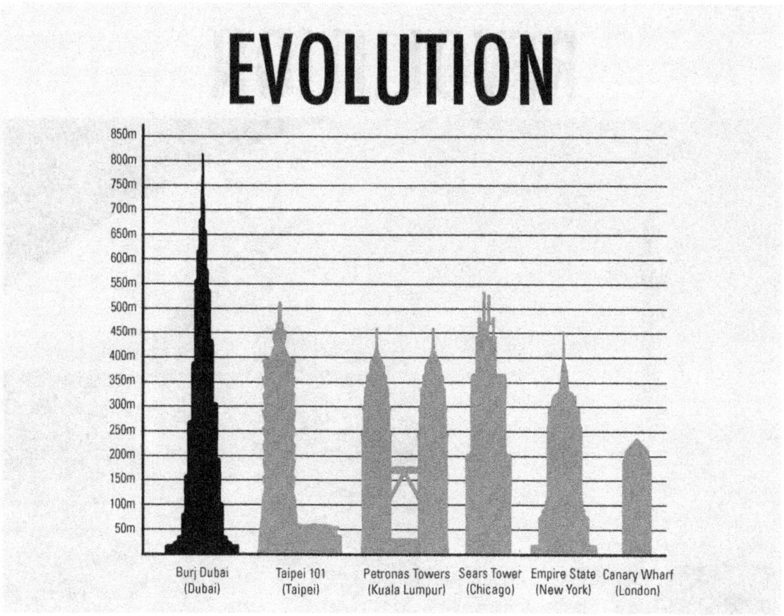

Google X, and they talk about the car. So most of the companies out there are saying: "How can we get an extra 10 miles to the gallon?" But if you come along, and you say, "I want a car which can run 10,000 miles to one gallon," then you need to reinvent what that car looks like. You need to start from scratch. And this is the idea of revolution. It is not taking something and improving it; it is starting from scratch and rethinking the whole process.

Keeping in line with skyscrapers, this is one concept that is changing the idea of what a skyscraper is. This is the ***Analemma Tower***, which was dreamed up by Clouds Architect Office in New York. The whole idea is to pull an asteroid into geosynchronous orbit around the Earth, tie some 50,000 kilometre solid steel cables around it, and then hang a 32 kilometre building from it. Now, obviously, this is concept, and it is not going to be real life anytime soon, if ever, but it is just that way of thinking. How can you think differently from the way everybody else has been thinking? It is such a cool idea.

REVOLUTION

When we look backwards, real estate has pretty much always been the same for over the last many centuries. Pretty much ever since the first caveman said to someone: "Hey, you can sleep in my cave if you give me a piece of your chicken." Real estate has almost always been the same. But all of that is now starting to change and that's why this is so exciting. We are literally at the beginning of five waves that are going to be coming together within the real estate industry into what I call "The Convergence."

So, what is the convergence? Well, if we take the word literally, it is when two things come together to form a new whole. For example, back in the day when the internet was invented, you had the internet and you had banking, now you have internet banking. Internet banking is a whole new thing. It's a thing by itself. You then had the same with mobile payments.

We are going to be creating whole new ways of doing things that, in 20 or 30 years time, our children, our grandchildren, would not even think of doing anything any other way, because it is just so natural for them. It just comes naturally because it would be so stupid to do it any other way.

We are really going to be seeing this happening within the real estate industry, and I will touch on this more at the end once we have had a look at all the different waves. When I talk through it, I am going to be talking about the waves as they stand by themselves, but the real power in everything we are talking about here is about the convergence of how different technologies and different trends within the different waves all come together as one. It is about taking pieces from this wave, taking pieces from that wave, pushing them together to create a whole new way of doing things that nobody has ever thought about doing before.

We are almost at the point within the property industry, that the internet revolution was in in the late 90s and, to me, that is so exciting. We are at a point where almost anything is possible. In fact, within the next two decades the amount of wealth that changes hands through real estate is going to be incredible. Real estate is the biggest industry in the world. It is almost double the global GDP, and all of that is about to get disrupted and start changing hands, become democratised, allowing more people to get into real estate and having a roof over their head, investing in real estate, receiving great returns, and that is so awesome.

To make things easy for you, I have created a rating system so that you can know when we are likely to see many of these trends and technologies coming in and affecting the real estate industry.

The very first one that we are going to be looking at is the ***immediately leverageable*** rating. Which means the things that are already in play, that we are already seeing coming through and that thing where people are already building massive businesses out of. This is really helpful for you to see a number of things: one, which directions some things are going, and two, to give you an idea of what you could be doing today to pivot your business, to add onto your business, to experiment and potentially magnify the results that you have been achieving.

The second one is the ***foreseeable future*** rating. This is very much that two to five-year mark. What are the trends and technologies that we are going to see come together in the next two to five years-ish? Now, this is really important, because this is what is going to really give you that unfair advantage if you were to create a business around it.

The story of Siri is the absolute best example of how important it can be to look at that two to five-year mark. You probably know Siri. Siri is the voice assistant on iPhones. Now, when the creators made it, they knew the technology to actually finish it would not be around for another three years' time and so, they started their company knowing that they would not actually have their finished product for another three years when the technology would become available. As soon as that technology became available, they were instantly out with Siri and Apple bought it straight away for a price of over $200 million.

You can be doing the same. You can be having a look at some of these technologies and trends knowing that in two to five years time, this particular technology is going to be available and if you were to leverage it in the right way, you could immediately be a market leader. The thing is, if you wait until the point where the technology is available, you are going to be starting from scratch. You are going to be starting from the base point along with everybody else in your industry who is going: "Oh, it is now available. Let's get in, play with it and see where we can go with it." You are going to have so much competition whereas if you know it is going to be there and you start planning from here, by the time you hit it, you are hitting the ground running.

You might even just be looking to implement it into your own business because you know it will make everything you are doing easier. You might currently be researching a large investment into infrastructure within your business and you know that in three years time, that infrastructure is going to become irrelevant. You might save yourself thousands, tens of thousands, hundreds of thousands or millions, who knows, by not implementing that structure right now. Instead, just simply make do with what you have for the time being because in three years' time, there is going to be something better out there that you can leverage and will do the job a lot easier – a highly likely future given where artificial intelligence is heading.

Next up, we have ***the midterm future***. It is kind of that five to 10-year mark. The reason this is so important is this is what is going to disrupt your business. At the moment, we are seeing it with many of the disruptors like the Airbnbs and the Ubers and they have generally been around for that seven to 10-year mark. They have been really popular for maybe five years but there are ideas starting to come through now that if Uber and Airbnb do not pick up their act, they could be disrupted as well. Even if you are the disruptor. PayPal is a perfect example. I do not know anyone who really likes PayPal anymore because they stopped innovating. They solved one problem. They stopped innovating and now, they are getting disrupted by other financial technology companies.

If you have been sitting there, if you are passive, if you have got a nice portfolio of properties, these guys are the ones who are going to disrupt you. These technologies and trends in that five to 10-year mark are still close enough to impact you immediately, and you should be thinking about what you can do. Yet, they are far away enough that they are not really at the forefront of your mind. It is not something you are thinking about every day because you are not going to see the effect from it for a while, so these guys are really important to watch out for.

Then, finally, we get to the ***distant future*** rating. The distant future is in that 10-year plus range. For the distant technologies that I am talking about, they are mostly going to be in that 10 to 20-year range. There is not too much that I can predict right now, nor would it really be that beneficial – unless you are doing a massive development – to look at that 20-year out range because the things that are going to be happening, the convergence that is going to be happening, that is going to create a whole heap of new waves that we cannot even predict right now.

These trends are important, so that you can know what is going to happen, to know what the far future is bringing so that you can create a destination. You can create your blueprint of where you want your company to go. For example, I told you right at the beginning of this book, that my life is dedicated to ending homelessness. All my businesses are dedicated to ending homelessness and we are planning on doing that through demonetisation of houses. Houses which are going to be free to live in, produce their own food, collect their own water and make their own electricity.

The aim of writing it in this style, is so that you can see which technologies and trends you can be immediately implementing into your business, whilst knowing what other tech or trends will be coming in the future so you adapt them into your business, or at least not be surprised when they show up.

One thing is for sure, with all the trends and technologies we are about to cover, you need to open your mind and reimagine what we consider the definition of a house to be!

CHAPTER TWO

LOCATION, LOCATION, LOCATION?

This first wave is probably the one that is really in your face, and most applicable right now. This is location, location, location. You have probably heard that phrase so many times as a property investor, and in the past, it's always been true. The right location is absolutely key.

But what if it wasn't? What if location no longer mattered because certain technologies emerged that almost made location irrelevant? What if it made very little difference whether you lived in an inner city suburb, or if you lived 100 kilometres away in the countryside on your own little piece of land?

These are some of the things that you need to be asking yourself as we go through this wave. Is location still relevant? Will it make an impact on house prices whether someone can commute to the city from 100 kilometres away in the same time it takes someone to commute to the city today if they were just 20 kilometres away?

So let's jump in and see what are some of the technologies and trends emerging that make the location aspect of property investing irrelevant, meaning location, location, location may not be so important in the coming 10 to 20 years than it has been in the past.

Globalisation

The first trend we are really going to focus on in this location, location, location wave is globalisation. I mean, you no longer need to live in the same place that you make your money. You can choose to live anywhere in the world, meaning people are suddenly discovering they can live somewhere else for a lot cheaper and live a much nicer lifestyle, in which case it does not make sense for them to be paying exorbitant city rents. Instead, they can go and live by the beach or in the mountains, where they can make the same amount of money, if not more, or they can choose to make less and have more time simply because they do not need as much money.

We are seeing a massive shift as people decide to live and invest all over the world. I am your prime example. I was born and bred in New Zealand, but if you ask me where I come from, that question is really difficult to answer. I have lived in Australia. I have lived in Europe. I now live in Bali. Are you asking for my nationality, or where I call home? Interestingly enough, what I pay for a seven-bedroom villa here in Bali on 3,000 square metres with my own pool surrounded by rice paddies is exactly what one of my tenants would pay in a South Auckland four-bedroom property. That really starts to put it in perspective of how, once you start thinking global, (and trust me, millennials are doing this) you do not need to live in one place anymore.

The internet is becoming better and faster, meaning people do not need to go to an office to work, and in fact, when virtual reality becomes mainstream, sitting in a virtual reality meeting could be just as realistic as sitting in the office in real life. This all begs the question of, do you even need to? We are already seeing in Auckland, for example, many CEOs moving out to more lifestyle areas, particularly places like Piha, working from home for the majority of the time, but it is still close enough to commute into the city, to attend a meeting, or to just check on staff one or two days a week. There is a massive trend of going global where anything is possible, and once you start thinking global, you will be able to find so many more opportunities that arise simply from this trend of globalisation.

Digital Nomads

Now, obviously, we cannot talk about globalisation without mentioning the rise of the digital nomad. Digital nomads are something that have been around for the last five years or so. Generally, they are going to be people in that sort of 22 to 35-year-old range, but it does span across all age ranges. Digital nomads are people who can live and work from anywhere, literally the laptop lifestyle. What we have started to find is that these digital nomads tend to move around a lot, hence the "nomad" part, and they just travel full time. Some areas are already taking advantage of this and digital nomad 'hot spots' are popping up all around the world from here in Bali to Medellín in Colombia and to Hungary in Europe. Digital nomads are flocking to places with a great lifestyle, cheap living costs and kickass internet. Of course I obviously cannot talk about digital nomads without mentioning Chiang Mai in northern Thailand.The ultimate Digital Nomad mecca. Digital nomads are mostly freelancers or contractors. Some people are remote employees. Others are entrepreneurs, and they just travel the world as they like. Generally, looking for somewhere cheap where they can be warm, have fun and live a great lifestyle on a very small dime.

These are people who have chosen lifestyle over work, because they have discovered, by living somewhere like that, their work-life balance becomes so much better, but not only that, by starting to be global, digital nomads are really starting to discover that they can figure out a few ways around different tax systems. It allows you to get a much better understanding of different cultures, and travel is probably one of the best ways to overcome prejudice. Digital nomads are probably what I would consider to be true global citizens, where you are not really from anywhere or a resident of anywhere because you are constantly on the move.

Now, this opens up many opportunities within the investment area. The first and foremost is the rise of medium-term accommodation. If you look back 10 to 15 years ago, you had two types accommodation. You either got a long-term lease on a property, and generally that property was empty, you would have to move all your own furniture in, you would stay there until the end of your lease or until you had to move, and you would go find another property, and again, you would be there long-term.

The other one was short-term accommodation, and this space was generally filled by hotels, homestays, backpackers, that side of things. In fact, if you were to travel the world and you wanted to live somewhere for that medium-term range, there is a good chance you are probably going to be staying in a backpacker's hostel because there has never really been such a thing as medium-term accommodation until the last five or so years, where people are coming only expecting to stay for two weeks to six months, without a contract in place.

This has become something that I call: housing as a service. They come in wanting a fully furnished room from sheets to cutlery, because, let's face it, you do not want to be taking all of that around the world with you, especially if you are travelling every couple of weeks or months. We have actually been doing this here in Bali ourselves, and I am going to talk about this later when we get to the coliving side, but I mentioned my villa before. I have opened this up as medium-term accommodation to other travellers, to entrepreneurs, to freelancers, so that I can live with other amazing people and build my community – another point I will touch on later.

With globalisation comes many opportunities for the astute investor. One possibility is to transform a large property into a medium-term accommodation service, such as I have done in Bali. Maybe you want to go live a global lifestyle yourself and become a digital nomad/nomad investor, or maybe you can see other opportunities inside this trend that I am yet to see – in which case – go and experiment.

Self-Driving Cars

One of the biggest impacts on how our cities are shaped, how we live, how we commute, how we travel, is the advent of self-driving cars. Obviously today there are aspects of automation within that industry, and Tesla is very much leading the way, but we can expect within about the five year-mark to see a lot more on the roads. We will not see massive change until the 10-year mark, but we will definitely be seeing some results within the next two to five years.

What exactly will automation and self-driving cars do for us? Other than completely impact the car sales industry, as well as other auto industries such as insurance and mechanics, we are already starting to see a trend towards transport as a service. This, obviously, made largely due to Uber, who created the ability to just simply order a car whenever you wanted one. This is going to become mainstream. In fact, in the future you are not even going to own a car. You are not going to own a car because you are not going to need to.

If you think about it, at the moment your car sits empty. It just sits in the driveway or a carpark for 95% of the time. That asset could be best put to other uses. Tesla has indicated that they expect car ownership and usage to look a little something like this:

There will generally be three options:

1. You own your own self driving car outright – there will be very few people who take this option.

2. You own your own self driving car, however, when you are not using it, the car is submitted to a rideshare pool, where it can go out, drive around, pick up other people, and then it becomes an income-producing asset.

3. You are just the user. Anytime you want to go anywhere, you just jump on your phone or whatever smart device we have at that time, order a ride and that will take you to where you want to go.

Let's think about some of the implications this is going to have on real estate. The first, obviously, garages are going to become completely useless. Although in saying that, when was the last time you actually saw a garage which had a car parked in it anyway? Garages are for storage. It is for junk. It is not for storing cars. In the future, we will not even need them. That means you might be able to do a garage conversion, you might be able to do something else smart and use the space better. It also means if you are doing a subdivision, you won't need to allow for a garage or car parking so some land which cannot be subdivided today, will be opened up and become developable.

Speaking of car parking, that is all going to become irrelevant too. If you think about how many buildings or how much space in a city is dedicated towards car parking, and then you think about the possibilities if it was all made usable again. For example, in Los Angeles, 14% of the land is used solely for car parks. If you can repurpose that into residential property or something more valuable, what will that mean for cities?

We have seen a trend of cities becoming more pedestrian-friendly, taking away car parks, taking away roading, and we will probably continue to see the same heading into the future. For example, we won't need street parking. We won't need the massive car parks that you find at malls. What then could be done there? Would that get developed further into other shops? Would it get turned into residential tower blocks? Would it get turned back into parks?

How it is going to eventually look is we are going to end up with a parking belt – much in the way that most cities have what they call the 'greenbelt' now – there is going to be little areas specifically for the self-driving cars to come back and park, get maintained and charge (because all cars are electric of course). Almost like little self-driving car transportation hubs. They are going to be located in close proximity to the busiest centres where people are going to need them most. Probably within a

10 minutes drive or so, but also far enough away that the land is cheap, it does not impact the neighbours and that side of things. Very much like the industrial areas that are around today. What is interesting is these are going to be maintenance hubs, they are going to be charging ports, they are going to be car parks. They are really going to be a multi-use environment for cars. In fact, if you own a parking building in the city fringe area, this could be the future that you start looking at going down. Providing the necessary infrastructure for a self-driving car hub to operate out of your building.

What we are also going to see is a lot more urban sprawl because suddenly commuting does not really matter. At the moment cars are very much something you do yourself and they are very purpose-driven. They are there to get you from point A to point B. As the technology progresses, what we define as a car is very much going to change because we are not going to be the ones driving it. Meaning, we are going to see a lot of innovation with layouts and designs of cars. Suddenly, you are going to have the ability to almost have a moving office, where you can climb into the car and do work on your way to the office. You might be able to have some sort of relaxation booth and you will have cars which will be able to have meetings in. It might even be possible that you can do a workout in the car on your way to the office.

Once we have self-driving cars, you do not necessarily need to live really close to the office. So again, people are probably going to move out, way out, and commute in. Maybe with the first hour they work in the car, arriving in the office. Or back to what I was talking about with the executives in Piha. They might be living somewhere like that. It is only a 40-minute commute and by the time you take out traffic lights and stop signs, these will become redundant because the cars are going to speak to each other, the commute is going to be way faster anyway. Cars are going to be able to travel faster, it is going to be quicker to get around the cities.

So what might be a 40-minute commute for you now, might give you the ability to move another 20, 30 kilometres away from the suburbs, have a nice patch of land, a little bit of acreage and still have the same amount of commuting time. It is just what you are doing on your commute that is going to change. No longer will you be sitting in a traffic jam, tooting, being annoyed. You might be reading the paper (not a physical one of course, get with the times), it might be doing some work on the laptop, having a meeting, making calls.

So, self-driving cars is definitely one technology that you should be keeping an eye on, especially once we get to the point where fully autonomous cars are legally allowed on our roads. Commutes will speed up, parking will disappear, garages will become redundant and cities will start to expand as people choose to live further away now that they do not have the hassle of driving to work each morning.

Drones

Another technology which is really starting to come into play and is about to turn exponential is drone technology. Now, you have probably seen drones about already. At the moment they're very much just a plaything, or else awesome for some photography enthusiasts.

In fact drones are already quite prominent in the real estate industry where photographers will use them to take photos of land, particularly bigger sections, or to get a unique perspective if need be. If you are a real estate agent or a photographer, that is one direction you could go immediately but, to be honest, that is kind of stock standard. There are already a lot of drone photographers in the market.

However, what I want to talk about today is more the direction in which drones are going to take housing, how we live, the shape of houses, and what they look like.

Because drones operate near to the ground, and even once autonomous, they are still going to be operating within the 0 to 100 metre high air space. Therefore, we are going to see clear skies open up. Not because of drones, obviously. They are still going to be up there, but we are going to start to see a lot of things like power lines coming down and being grounded or disappearing altogether once we get true wireless electricity. But, really, all the things that are out there cluttering our skies at the moment are going to disappear. So it is going to be a lot more beautiful in that perspective.

The other thing to disappear is going to be the letterbox. Now, depending on where you live, it might be different. In New Zealand and Australia, we very much have the traditional letterbox out near the road. In England and other European countries, it is kind of built into your front door. Obviously, if you are in an apartment, then you have the massive row of letterboxes down in the lobby. They are all going to disappear because the drone's just going to fly it in and drop it straight on some sort of drone landing pad.

Drone deliveries are in experimentation at the moment, with many things being trialled, including the first drone-delivered pizza. Amazon are doing big trials in England, with the aim of extending their drone delivery service across the world once they have perfected the technology and received permission from governments. This will all happen within that two to five year range. So expect it to happen pretty soon.

In fact, Amazon is really leading the way in drones at the moment and, in particular, drone deliveries. They have already applied for a number of patents. The most important for what we are talking about today is a very advanced drone landing pad, a sort of drone letterbox. It is going to have its own unique emitter, a radar or a unique code that it emits, so that the drone can find it and deliver the package to your particular address. I guess you can compare it to an IP address on a computer.

So, the drone will land and it will place the package on this landing pad, which will have an inbuilt trapdoor feature. There are many different add-ons, which have all been applied for within the patent, from temperature control, to pest eradication, and obviously water penetration. But this is very much one area that drones are going to play a massive part in, removing the need for letterboxes.

If you want to start thinking about how this could be applied, because Amazon are leading the way in that, it would be more on the installation side. Who are the people who are going to want these installed? Can you set up an installation company? Can you become a reseller or a re-marketer? Again, because it is two to five years away, it is definitely something that you should be keeping an eye on.

The other aspect of how drones are really going to get in and start disrupting things in the real estate industry is very much in that commercial space. There are a lot of companies which all revolve around transportation and logistics. Obviously, they are going to get disrupted, but in the same way self-driving cars are going to create parking hubs and parking belts around the city. We will also see drone distribution centres in a similar style 'industrial belt' complete with drone landing pads, charging pads, mechanics for drones, that side of things.

We will probably see drone companies go either one of two ways. If a company is going to do deliveries, then they could buy their own drones, in the same way that a pizza shop has their own employees to deliver pizza. We might see it that way, where the company owns drones to go out and deliver their product. Or, we will see something different, which will be like UberEats – Thuisbezorgd if you are in the Netherlands, GO-JEK if you are in Indonesia. But, I think most people understand UberEats, where it is now a transport company that owns all the drones. Whenever something needs to be delivered, they would just press the button, the drone would come, pick up the package, it automatically knows who to go to, it drops it off, and then it goes to the next delivery nearby.

My best guess is that while some people would own their own drones for delivery, we will most likely see the drone deliveries as a service, and that is going to really disrupt many of the commercial belts and distribution companies.

Again, Amazon has filed for another patent to have a massive airship, up in the air, and that will almost become a fulfillment centre, so it can move around depending on what is going on. If there is a big game on at the stadium, the airship can sit above it and deliver everything from pizza, probably not beer because of age requirements, but merchandise, that side of things, down to the people on the ground. And, we will probably see this in different areas of the cities, where a massive distribution airship sits up in the clouds and the drones simply go back and forth. This has been highly praised due to the energy efficiency for the drone, because it flies horizontal and then just drops up and down, so obviously it is using a lot of gravity, which means that it preserves its energy.

That is probably much more in that 10, 15, 20-year space, if it ever does come into play, but it is just another really unique outlook. It also means that if everything goes into the sky, then we can expect a lot of industrial real estate to go out of business. We already know we are going to see a lot of commercial real estate go the same way, and if you can buy something online and have it delivered to you faster than it takes for you to go to the shop and even look for it, then there will be a lot more retail companies experiencing hardship. If you are a commercial investor in any of those areas, you want to start looking at how you can do things on a more experiential basis, rather than just straight out selling goods.

Flying Cars

The convergence of drone technology and self-driving cars has led to one of the most exciting technologies in this book: flying cars, and here is where things get crazy. It is where science-fiction becomes science reality. Because of drone technology, we have existing flying cars today. It is only a matter of 10 years or so before they become mainstream. In fact, in Dubai, they are already testing flying cars. How cool is that? Now, at the moment, they are not allowing human travel in the flying cars, but they do have autonomous cars flying around the city that once they have finished testing and they have sorted out the regulations and rules and that side of things, we are going to be able to fly. The future is here now.

A Chinese company created the Ehang 184, basically a flying drone car, which is completely autonomous and can hold one person. Since then, there are a few other players who have entered the game. It is all using vertical takeoff and landing techniques. People have been playing with flying cars for years. But, this is pretty much becoming reality now – although they are not going to become mainstream for another five to 10 years and will be pretty much dominating globally as a legitimate form of transport in about 20 to 25 years.

Uber Elevate has its own little department where they are working to make flying cars part of their business model, in the same way they turned cars or transport into a service. Flying cars will just be one of those transportation methods. Again, with your phone or whatever device we have in the future, press a button, in comes a flying car, and picks you up. You hop in, you do not need to drive or control the vehicle, it knows exactly where you are going and it will just take you there.

Now, they have made projections for what is going to be possible, and as you can see on this diagram here, to go from San Francisco to downtown San Jose, traditionally, if you wanted to go by car, it would take you around about two hours and 12 minutes. However, it is only going to be 15 minutes by the time you take a flying Uber. Price-wise, in the long term, it is going to be quite a lot cheaper than any other method.

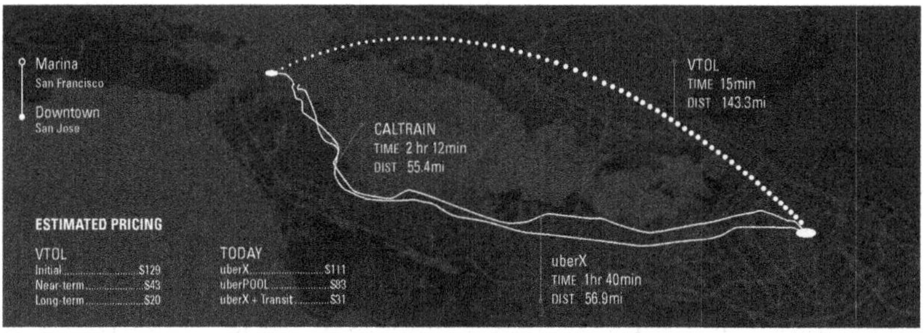

As I said, Dubai is already testing flying cars. If we take a look at a map of Dubai, the furthest point in the emirate is only around 65 km from the actual city centre. Meaning you can almost get anywhere within Dubai by flying a car within 10 minutes. It completely shrinks down the way we live, as well as accessibility within the city. It dramatically cuts down the transportation time it takes to get anywhere or to ship anything. In the same way that the world is becoming a lot smaller, cities and countries are going to become a lot smaller and in essence you will be able to get anywhere a lot quicker.

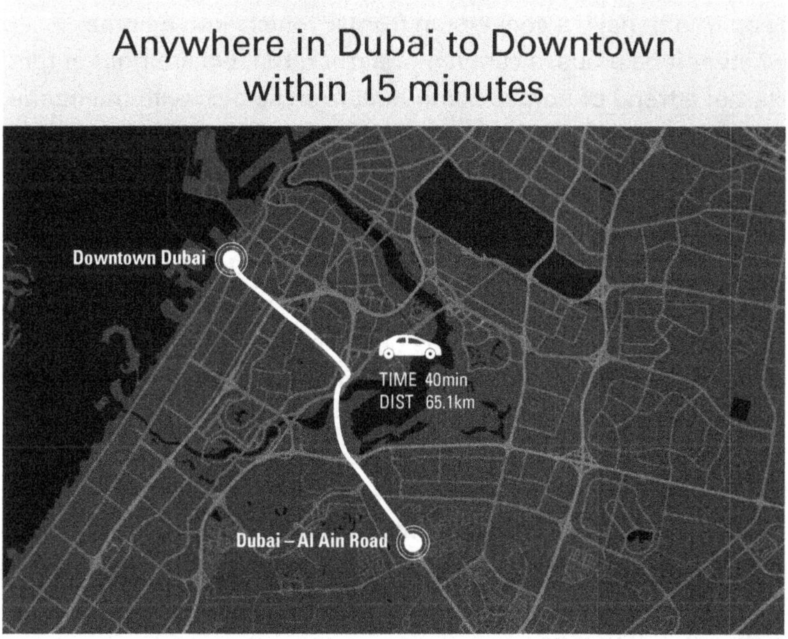

Once this comes into play, location, location, location does not necessarily matter so much because, between self-driving cars and flying cars, you can be anywhere within 10 to 20 minutes. Again, looking at Dubai, if you can be anywhere in the emirate within that 10-minute range, then it doesn't really matter where you live. Now, you are basically just living where you want to have a nice view. Self-driving cars, flying cars, these are really going to be, I think, the biggest changers in the layout of our cities and what we even consider to be a city.

One thing future city planners and governments are going to have to think about is – when everyone has access to flying cars – what will that mean for border security, particularly in border towns or countries that have land borders with other countries? Smuggling will become a lot easier (be it people, drugs or goods) and many smaller, poorer countries might not have the resources to police the skies.

Transportable/Mobile Homes

While mobile homes are not a new concept, they have always been traditionally a conversion from a vehicle into a home, whether it be a bus, a caravan or a truck. But, we are right at the start of a trend of homes being built from scratch with the intention of them being mobile right from the get-go. This is going to lead to a new type of nomad, with families moving around their own country every so often, taking their entire home with them.

Already concepts are being created from floating homes, as is the case of the Unidentified Floating Object **www.jetcapsule.com/UFO/UFO.html** which has been designed from scratch as a floating residence. It includes underwater bedrooms and has the ability to connect in with other UFOs to create a neighborhood flotilla. Back on dry land, we are seeing a range of mobile houses which unfold from the back of the truck, creating a mobile two-storey house, that fits within trailer specifications, and requires no permit. Simply drive up, hit the switch and the entire house unfolds before your eyes. Plug in the power and water and off you go.

Another concept is a transportable module that can be put on the back of a truck and delivered anywhere in the country, off-loaded and then reloaded any time you wish to move on to the next location.

I believe this is going to see a rise of people doing medium-term stays in different locations around their own country. Drive to one town, spend two to three months living there before moving to the next location for an equal amount of time. This will open up an opportunity for property entrepreneurs to set up community grounds for these types of nomads, much the same as the campgrounds we have today. These community grounds will be set up with power and water that can be plugged into the home as well as having a communal living area to engage the community. Tenants will simply pay for the time they stay in that location and any utilities they use, before the truck picks up their home and takes them to their next destination.

This trend might even merge with self-driving car technology and we will see self-driving houses on the road, as well. Already, RV companies are experimenting with the technology, but once it becomes more refined and established, it will not be a far cry to envision a day where we leave the nightclub at 3 am, push a button on your phone for your house to come pick you up, sleep the night away on the road and then wake up on a remote beach where the fresh ocean air will help cure your hangover, while the house, aware you have woken, has already started brewing a nice cup of coffee for you.

Super High Definition

At the moment, we already have high definition TV. You can usually tell because the actors have to wear so much more makeup to hide imperfections. Actors, who once upon a time would have looked attractive, now appear less so in high definition. However, very soon we are going to have super high definition, where it is so clear that our eyes cannot differentiate between the screen and real life.

In the past section where we were looking at flying cars, I mentioned the only reason you might want to live in a particular area or pay more to live in particular area is because of the view. High definition or super high definition removes that scarcity. I can literally have a window, which has some sort of screen embedded into it, and if I want to look at the Eiffel Tower, I can simply set my view to the Eiffel Tower. My view of the Eiffel Tower would be more realistic, and my eyes would not be able to tell the difference than if I was literally standing in Paris looking at it.

So, I can wake up each day and if I want to wake up in Paris, if I want to wake up in Thailand, if I want to wake up on my own private island, I can look out my window and see all that in super high definition. So, it takes away the scarcity factor when everybody can have the same view, and have that lifelike real view. But, what it also does, is open the doors for affordable housing in bulk or areas that are a lot less desirable.Super high definition makes every area more aesthetically pleasing.

For example: There is a concept that has been dubbed an 'earth scraper', and designed for the centre of Mexico City. Now, in Mexico City, they have a law which prohibits the height of a building to be more than eight stories. One architect came up with this idea of an inverted pyramid, going fifty or so stories below ground, and then having a flat roof or bottom of the pyramid, which is right where the town square is.

Obviously, if you were to live 50 storeys below ground, most of the time that would be pretty horrible. It would be dark and it would be dingy, but if you were able to use super high definition to have windows, which look out over the ocean for example, that suddenly changes the ambiance of living in a place like that. It makes it somewhere that people would actually want to live. Tie it in with the design of the actual pyramid where the centre would be like a big hollow light lens and have lots of green spaces and little parks in there, to be honest, people might not even know the difference.

That is how realistic this super high definition could be. So, if you are looking, and again, remembering we are in that sort of two to five-year range until super high definition becomes a reality, I would not expect people to start implementing it into real estate until that probably five to 10-year range. But, it could be something that if you are looking at doing a development that doesn't have amazing views or it might be next to something that is not so pretty, simply by installing super high definition screens where the windows allow you to provide a better view, a better ambience, and maybe it could even mean that you can charge more of a premium for the 'view'.

Hyperloop

As we start to look at the decade-plus area of location, location, location, we are going to come across the Hyperloop. When I was in Los Angeles, I met one of the engineers working on the Hyperloop. She had come from working for NASA. Seriously, she was like The Big Bang Theory in real life. I remember we were at a cocktail party, and I asked her a question about how the Hyperloop would work, or that side of things. She was like: "That is a very good question. I have not thought about that." She was like, "I need to go work this out" and left the party. She wanted to go grab a pen and paper and figure out whatever it was I had asked her. It was so funny. Unfortunately, I never actually got her contact info so I never found out the answer.

Anyway, I digress. The best way to explain the Hyperloop is as a magnetically levitating personal train carriage that speeds down a vacuum tube.

What will the Hyperloop mean? For those of you familiar with New Zealand, it pretty much means that you can travel from Auckland to Wellington, CBD to CBD, in approximately 40 minutes. That is crazy. It takes longer than 40 minutes to go from Auckland CBD to the airport, let alone having to also be an hour early, to then do the hour flight, and then get from Wellington airport to Wellington CBD. You can see that this technology is really going to shake things up. It is tied in with self-driving cars, because obviously, who else but Elon Musk is behind the Hyperloop? But, the idea is that you had to have a self-driving car, drive into a Hyperloop pod, which would then enter the vacuum tube and travel frictionlessly at very, very fast speeds, and essentially be able to come out the other side, where you are still in your self driving car, heading towards your final destination.

Hyperloop have open-sourced everything they are learning and have a test site that they call the 'develoop' – a developmental Hyperloop which they are playing with in America. Even Richard Branson has come onboard, and now he has got a virgin Hyperloop that he is playing with. It is almost going to be like the Silk Road all over again when you start to look at some of the tracks that are available, and what is really possible. Hyperloop will probably play a bigger part in shipping food and goods more so than it will in transporting people. So, expect land prices around hyperloop hubs to increase in value dramatically.

By applying this technology to New Zealand, it means that if you can go from Auckland CBD to Wellington CBD within the space of 30-40 minutes, and you tie that in with flying cars and self-driving cars, it almost opens up the entire North Island to be within a one-hour commute of Auckland or Wellington CBDs, not to mention all the other towns and cities in between. That is just crazy. It completely redefines what we consider a city. It essentially means the entire North Island becomes one commutable city. This is where I talk about the disruption potentially coming in. Because you've got to ask yourself, if you can buy a house in Auckland for a million dollars, or you can go into one of the small rural towns and buy five acres and have a really nice house for $100,000, and still be within that one-hour commute from Auckland CBD, what are people going to be doing? It is going to become very much about lifestyle, as it is about where you are actually living.

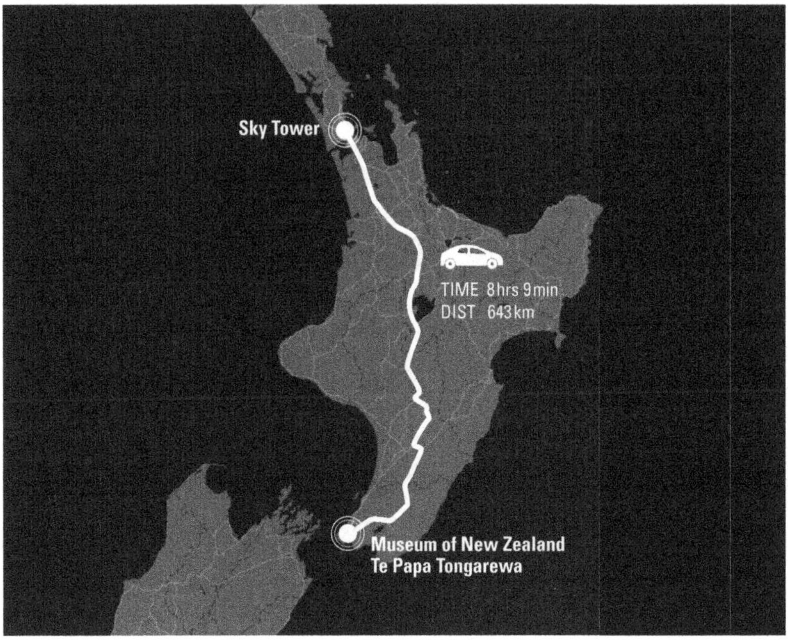

If we take a look at the South East Coast of Australia, in particular Sydney, and we draw what would be within an easy one-hour commute via Hyperloop from Sydney CBD, you can see that Melbourne and Brisbane fit within that area. Essentially, it becomes cities within cities. What is the definition of a city anymore? If you can live in the centre, and again, go into the middle of nowhere, but live within a one-hour commute of Melbourne, Sydney, and Brisbane, is that going to be an area that pops up as a new liveable area? Where you can buy dirt cheap land, build on it, and then commute to any of those three cities, as well as all the smaller cities, towns, and provinces within those areas. Canberra obviously falls within that area, as does Byron Bay, and many other towns. So then, what will that mean?

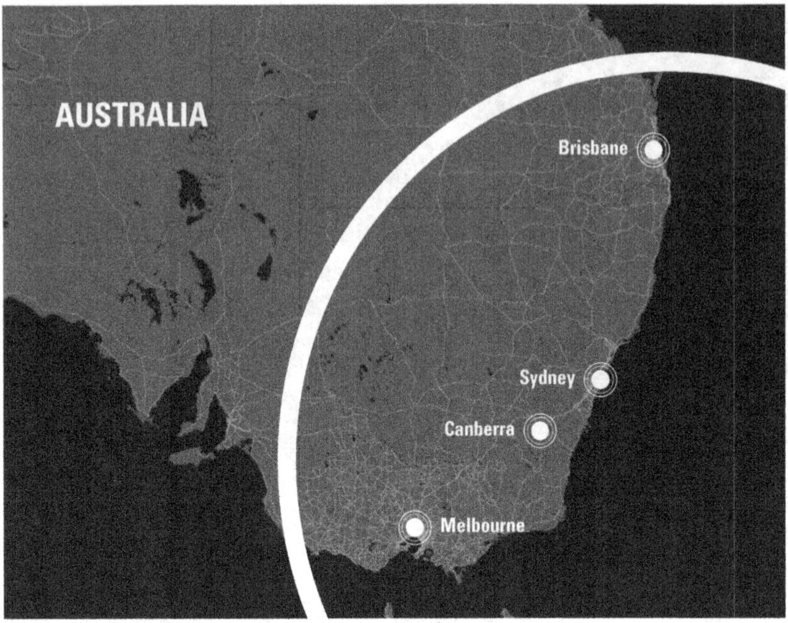

This brings us to the end of our location, location, location chapter and, as you can see, there are some technologies and trends that are really shaking things up.

We have seen how the distance from the city may not be as relevant in the future as it is today, simply because of better technologies that can transport people faster, safer and even change the way we commute. From trends such as globalisation that mean people don't even need to live and work in the same country, to how we can use super high definition to provide anyone with any view they desire and have it look more real than their eye can notice.

So, there are a few questions you need to ask yourself.

What does it mean if people do not have to live right where they work? If people can live anywhere. We have looked at flying cars, we have looked at self-driving cars, we have looked at Hyperloops. What will that do to real estate prices?

And, I do not know the answer to this, but will people still want to live in metropolitan areas or will there be a movement out to the suburbs? Will people continue living in large cities or will they go out to the country, living on a five acre block?

What will it mean when people can get anywhere very quickly, very cheaply? What will it mean if we do not need garages, if we do not need letterboxes? Are there opportunities there that you can take advantage of? How will any of this affect your current investments? How will it affect your business? Is there something that you can do to make an immediate pivot? Is there something that you can do to look at that two to five-year range so that you can tweak something within your business and make the plan of how you are going to get there so that when the technology becomes available, it is all set and boom? How will this impact what you are doing with some of your current investments? If you are in the commercial space, might you think about selling something before many of these things become mainstream?

The world of real estate is changing alright, and this has just been the first wave.

CHAPTER THREE

MY HOUSE MY CASTLE

So let's break into the second wave, My House, My Castle. This is where technology is coming through that is going to allow us to ultra-personalise our own homes. Even to the point where we will be able to have different art on the walls and be able to see different furniture in the house, than the other people that we are actually living inside the house with. Cool hey?

At the moment if you want an architecturally-designed house, or if you want something a little bit different, then you are going to be paying through the nose for it. However, technology is emerging that is going to allow us to have ultra-personalised houses so that we can put our own spin on things affordably, if not free.

We can almost create any shape or design of house we want. This will mean that architecturally-designed houses are possibly going to be worthless or at least worth less. And, it is going to be interesting to see how it comes out because at the moment house values are based on scarcity. So, if you think about, again an architecturally-designed house with an amazing view, and then you think about a stock standard cookie cutter approach to housing. The cookie cutter one is always going to be worth less.

You think about a suburb, which is full of cheap houses where all houses look the same because they want to make it affordable. It is going to be worth a lot less. However, within the next few decades, it is going to be possible to build housing just as cheaply as the cookie cutter approach now, actually probably a lot cheaper, and still have your own architectural style.

So, this is going to be an amazing wave where all of a sudden we can have ultra-personalised houses. Let's get into it and have a look at what technologies and trends are emerging in this space.

Recycled Materials

The first of these trends, which is happening now and can be actioned immediately, is the use of recycled materials in housing. It is the ability to go out and create your own home, upcycling materials using natural materials and recycled materials. This is really exciting because, obviously, there is a massive pollution problem with the planet and some people have figured out ways not only to clean up the planet, but to also provide shelter. And, for me, that touches me to the core.

Obviously, the most famous of these would be Earthships, which started around the 1970s. They use all sorts of methods to create a passive home that utilises solar energy, wind energy, thermal insulation, and has some sort of food aspect to it – whether it is a built-in greenhouse into the home, or even aquaponics. I have seen all sorts. It has its own self-contained sewage system, and sometimes even a water processing plant through a living wall or other aspect to it.

Earthships are really cool. The most common design is where they pack the earth into tyres, stack them, and then cover the outside with more dirt, almost like a hobbit hole sort of thing. And, these are all passive, so they use less energy than it takes to build them or to run them.

I really think in the future we are going to see a lot more recycling in buildings and a lot of 3D printing, which I am about to get into. The housing industry, particularly in China, is able to recycle concrete into the paste they 3D print with.

This is becoming a trend in a few countries where pollution and rubbish is a major problem and I have seen something come out of Columbia where one of the innovators was able to shrink plastic, melt it together into some kind of Lego blocks, and then build homes for the homeless and the poorer community. All of these homes are made out of plastic recycled blocks stacked on top of each other to create the shell that you need for the house.

Another one, which is really dear to my heart, is something happening out of Wellington, called Poly Palace. These environmental entrepreneurs have figured out how to recycle polystyrene and turn it into tiny homes. They are working with one of the local community organisations to build these tiny homes for some of the poorest and homeless people – such great work.

So, there is a lot going on in the environmentally-friendly space. Whether it is creating a truly passive home, or whether it is using recycled materials, the future is definitely going to be in, one way or another, something that makes less of an impact on the planet. So, again, this is something that is happening right now and no one's really commercialising it at the time of writing this book.

A simple way to utilise or leverage this trend is to start doing something in that passive home space. You can charge a premium, right, if people know that they are not going to pay such high electricity or water bills in the future.

Smart Houses

Smart technology is all the rage right now. We started with smartphones, and we have moved to smart watches and other wearable devices. In the future, we will probably even have smart contact lenses or even neurological chips implanted directly into our brains.

However, smart homes have started to really come out in the last five to 10 years. They are not really commercially mainstream right now, but they are starting to make a dent. At the moment, smart homes are very much on the energy efficiency side, so a house being able to know when someone's not home and will turn the lights off and can turn the thermostat down, as well.

But, for richer people who are looking for something really cool, there are some awesome features out there. From a bed which will gently wake you in the morning, to shutters, which again, instead of just an alarm clock buzzing they will slowly raise. Or, even better, if you have got smart windows, they will just turn from being pitch black to completely transparent over the space of about five minutes, allowing you to wake up a lot more naturally.

Another really cool aspect of smart houses which I think is just insane, is when you wake up in the middle of the night needing to go for a pee. Instead of having to turn all the lights on and waking up your spouse, the house can actually sense that you have woken up and you have gotten out of bed, and it will just turn on minimal lighting along the hallway until you get to the toilet. And then again, instead of having this blaring white light, it will just be dim enough so that you can see what you're doing and then it will turn off the lights as you walk back to bed.

Many cool apps like that are starting to come into the marketplace, which are allowing you to really interact with the house a lot more, rather than just having standard lights on, lights off.

We are at the point where your house is very much linked with your phone. For example, if you are coming home from work, your phone can recognise when you are a certain distance from home, send a message to the house, and say, "Hey, we are five kilometres out. Can you turn the oven on because we are going to be cooking dinner?" You can be at work, realise you have left the iron on, simply use your phone to turn the iron off at home. This is all part of the 'internet of things' where everything is going to be connected. They call it the '50 trillion sensor network' where everything is connected with each other, and they all know what is happening. We are still right at the bottom of the exponential curve when it comes to the internet of things. Seriously, we are so low on the curve that it leaves so much room for anyone who wants to jump in and take advantage of that technology. You have the ability to create a company around any 'dumb' home device and turn it smart, before connecting it in with all the other devices.

Smart homes are really cool, but it will take a while to see them become mainstream, I think probably within the next 10 years we will start to see certain aspects integrated into new developments. We will see devices rather than necessarily a whole brain of the house installed, for example. Two years ago I stayed in an Airbnb that had Nest, a device which controls the thermostat. The owners worked from 9 am to 5 pm, Monday through Friday, so the house knew that at nine o'clock it needed to turn the heating off. Then, at about 4:30 pm, it turns the heating back on before they get home so the house is not using unnecessary heating during the day, but is still a nice temperature for them when they get home.

Whether it is a single device or a full brain for the property, we going to see smart house aspects becoming more mainstream and over the coming decade, new developments, particularly will be adopting many forms of the technology. From the 10-year onwards mark, I would be expecting nearly every house to have some sort of smart device in it – just like how a lot of western countries now have smart metres for electricity.

Smart Cities

In the same way we are seeing the rise of the smart home, we are also seeing the same trend happen at a citywide level, where many cities are implementing smarter devices to control the city and run it in a much smarter manner. One such device is a sensor in rubbish bins, to know when a rubbish bin is full and needs to be emptied. Each morning, artificial intelligence will plot a course for the rubbish truck to take the most efficient route to empty only the bins that are needing to be emptied that day. Whereas at the moment, the rubbish men have to empty every single rubbish bin as they come across it. Even if it is half full, they still remove the plastic bag and replace it. Not only does that create more rubbish from the plastic bags they use, but it is a lot of inefficiency and wasted time. Meaning, the fullest rubbish bins are now overflowing because they're wasting their time in doing the empty rubbish bins.

Most western cities already have smart devices installed in one way or another. You may have even used one without knowing it. Smart traffic lights can measure how many cars are waiting, and then it decides on who to let go and who to stop. Even down to using sensors in water pipes, to be able to detect if there are any leaks and where that leak is, instead of having to dig up the whole pipe to find the leak.

Smart cities also take it to a whole new level, as well. A city is being built from scratch in South Korea called Songdo – called the connected city. It is so smart. It was designed specifically, and built from the ground up, on 70% reclaimed land. It is designed as an escape from nearby Seoul and is targeting those working business people that would traditionally live in Seoul, but are looking for more of a lifestyle change.

It doesn't even have rubbish bins because in your home there are suction pipes underneath the streets, so when you put something into the rubbish bin at home, it takes it out directly to the rubbish dump where it gets recycled or sorted, and then moved on from there. Everything is connected: from the home to the school, to the office to the shops, to the government so you always know what is going on. You can track buses and even track your kids.

Songdo has been created as this really well-thought-out city utilising all these latest technologies, and just making it so easy for people to live a really nice lifestyle. It is a lifestyle city. We are going to see many more cities such as Songdo pop up in the future, whether it is an entire city built up from scratch like Songdo was, or it is an existing city turning smart like Singapore is. Smart cities are definitely the way that cities around the world are going.

Augmented and Virtual Reality

So, as we move into the foreseeable future, one of the big technologies that is really going to play a part is virtual and augmented reality. Talking about "My House, My Castle" and ultra-personalisation, I think augmented reality, in particular, is going to play a massive part. Virtual reality no doubt will have some influence, and I will get onto that in the next wave, but very much being able to ultra-personalise your home with furniture that only you can see is absolutely, I think, mind-blowing.

First, augmented reality needs to become more mainstream and easier to use. At the moment, augmented reality is either looking at something through your phone or putting on a Hololens-type device. But, once we can get it down to the point where it is just a contact lens or some other compact device that allows you to see in augmented reality, then many more apps and uses will come into play. Augmented reality will then become a full-time digital layer over the physical world.

With AR, you could walk into a home and it could be painted in your own colours. You could look around and everybody in the home could paint a wall a different colour. I could look at it and say: "I want to paint this wall blue," my wife could look at it and say, "No, I wanted it pink," and my children could say, "Actually, no. We want to go for black with yellow graffitti across it."

That is one thing you are going to be able to do, and then you can virtually decorate your home. Even adding the Mona Lisa to your wall will not be a problem. I could just go to the virtual shop and buy a new print any time I wanted. I could have that changed every single day. I could have messages, reminders, all sorts of things painted up around the wall. One company that is doing this really well is Magic Leap, and I have included a link so that you can see some of the things they are doing: **www.magicleap.com**

One thing we can expect around the two to five-year mark, is digital shops selling digital furniture and digital artwork for your home. I would expect these digital homewares to be selling for very low sums and a whole new breed of designers will be produced. In the same way that people play Sims now, it is almost like you are going to be able to have your own SimCity in real life through your own eyes. Decorate how you want. You will be able to watch Netflix or TV or YouTube up on your wall with no purchase of a physical TV necessary. This is really going to remove a lot of the furniture that we have as it becomes redundant. If you think about it, from the speakers to the TV, to the DVD player or anything you have there, anything that is not used for physical comfort is probably going to be digitised and sold in some sort of digital store.

So, that is one space that someone reading this book could actually go through and implement – knowing that in two to five years' time this could be a very big mainstream business.

Another aspect that virtual and augmented reality are going to have is when it comes to doing a renovation. For example, if I wanted to paint my wall I could simply log in through virtual or augmented reality and the shop would pop up in front of my eyes. I could be standing in my home looking at the wall, simply changing the colours, trying to decide which one I liked, and then once I found one, the shop would deliver the paint or I could go to the shop and buy the paint.

Many aspects like this will soon be possible, if I wanted to try putting a window here or cutting a doorway here, taking measurements, all these aspects will be available utilising augmented reality. You can stand there designing everything, seeing it before it is even created. This will not only save a ton of money, but it makes things so much easier.

CASE STUDY

I hit up my friend Jon Masters who is paving the way with some interesting developments using the Hololens. I asked him how he thought AR will change the real estate industry, particularly given the technology and business model he is working on with **www.yougo.world** – his new startup allows anyone with a Hololens to become a guide and provide a live digital walkthrough of any property, attend a concert or visit a historic site, plus many more applications – Here is his response:

Hi Kyron,

*From a YouGo.World perspective we're most interested in the remote presence aspect, so our thinking has been along the lines of remote property inspection. As a landlord or prospective landlord, you'd connect up with a local 'YouGoGuide' who would do a live tour as directed by you. So, if you have a thing for inspecting attics for example ::**cough**:: you could do that 🤖 As I'm sure you already know, there are already "pre-recorded" VR tours being offered by estate agents but, as with photos, you're not necessarily going to get to see enough of what's important for you. I was recently looking at property in India for example, and I got really frustrated at the photos!*

For property development I think there are many more opportunities. For example, being able to use AR to project a development into the real space it's going to occupy. That could be a really useful way to support planning applications for example. Also, where property is going to be extended, it might be great to be able to visualise that before starting the build, perhaps as a sales tool for the builder/manufacturer. Other opportunities might include measurements of gaps etc so that you know that you're keeping to planning rules or you could use AR to visualise and model groundworks, view restrictions and other things that might be difficult to measure in elevation diagrams but could impact a planning application or the value of the property you're building.

For your mission of creating the chance for everyone to have a roof over their heads, I was thinking about how something like that could be delivered and how AR/VR might support that. I'm assuming that prefab / 3D printed housing is really attractive here because it can be built cheaply, delivered pretty much anywhere (especially if it's 3D printed) and is relatively easy to construct. I don't know much about the costs involved in a build but what I understand is that the bulk of the cost is labour. I'm assuming that giving communities the support they need to build for themselves would be what you're looking for (I may be wrong!) but imagine if the build instructions could be created in 3D AR. It would identify all the parts, show (not tell) the people exactly how to perform the build, teach them the assembly skills they need to get the job done and walk them through the entire process. The community are then left with a huge range of skills that could help them going forward.

Just a thought!
Anyway, hope that helps 🤖
Cheers, Jon

3D Printed Houses

One of the technologies which I find most exciting, and is definitely going to disrupt the construction industry, is 3D printing. 3D printing was invented back in the '70s. However, it is only really become commercially viable and mainstream within the last five years or so, and in that time, we have seen many companies start playing in the 3D printing house industry. To be honest, no-one has done it really well just yet, but there is a competition. It is like a new arms race where we are seeing America, who were the pioneers of the technology. We are seeing Russia, who are probably doing it the best at the moment. We are seeing China, who are 3D printing panels and then putting them together. So, not quite what I would consider true 3D printing, and we are seeing Israel coming through with some of their own technologies as well. And, they are all in this arms race of who is going to be the first one to 3D print a house and to really come through with a mainstream 3D printer.

Someone else I should throw into the mix there is The Netherlands. They have already 3D printed a house. They did it a little bit differently though, because they were more focused on creating sustainable materials and playing with the printing of different parts than actually going out to try to 3D print a whole house. I believe the Russians are the ones leading the way, however, with their crane-type 3D printer, which just pumps out the concrete. It is all digitised to be able to print layer, upon layer, upon layer, and then at the end of it, have a full house.

3D printing is drastically reducing the amount of time it takes to build a house, from months down to hours – typically 24 hours to maybe 48, depending on how big the actual house is, and there is only room to improve. The biggest impact on the construction industry and anyone who owns a construction business is that there is very little labour involved Three people can build a house in 24 hours. It is zero waste, because there is no need for offcuts; you are only layering the amount of material that you need. And then, also, safety wise, the construction industry is one of the most dangerous industries out there, and so there are so many injuries all the time from power tools to even the simple hammer, to the amount of rubbish lying around creating trip hazards, and it eliminates all of that.

So, we are going to see a massive change in how people build, the speed with which we can build, you can have a house go up overnight, and then also the shapes that we can construct. At the moment, to build a house which is not square, you are talking about a big expense. You will need to bring in architects and all sorts of specialists, yet once 3D printing becomes mainstream, it will cost the same to print a curve or to print a star as it does to simply print four straight lines. We are going to see a lot more personalised houses coming out, where people can simply print whatever shape they want.

Once 3D printing becomes more established, a business opportunity will open up for someone who wanted to create a marketplace for people to either design their own homes or to buy existing plans. Simply put, in the same way that you have 3D printer blueprints for everyday household items and toys, you could get a 3D printer blueprint for your home and simply call up the 'printing' company. They would come over with the tools, technology, and then simply upload your print schematics, and the 3D printer would do the work from there. It's really exciting, because it will completely demonetise the whole construction process and should allow us to build a lot cheaper homes.

In Africa, some organisations are playing with 3D printers which are building houses out of mud. Obviously, Africa has a lot of history of having mud and cow dung homes, so it can 3D print these houses out of mud, meaning it is even cheaper again, because it is all naturally abundant resources. It is just then taking the cost of the 3D printer, which is not going to be much in the long run, particularly because the service charges in some African nations are cheaper.

What is really interesting is looking at some of the business models coming out within the 3D printing companies, where it looks like many of those businesses are going to license the technology rather than actually sell the 3D printers. This really goes in line with major changes we are seeing in how businesses are thinking about and monetising their products nowadays. In the past, they would have just had a production line, simply selling the 3D printers, and then having to go out and find a whole heap of new clients, and new purchasers. Nowadays, it is almost like 3D printing as a service, where they provide the 3D printer and then you just pay a licensing fee for every house you print.

I think 3D printing, is going to be one of the biggest disruptors in the construction industry in the next 20 years, so keep an eye on it.

As the world becomes more connected, we are seeing homes become more connected as well, not just to each other, but also to ourselves, offering unique experiences to every user. I guess in some ways, you could say that the home is becoming the next user interface. As the world moved into the latest decade, we saw a move of business models from mass production and marketing to personalisation for the user – this is soon set to move into the real estate industry on a massive scale, which is going to be a whole range of new questions to be asked, about how any of these technologies will change how we live and do business.

Could something listed in this chapter put you out of business? Or is there something here that you could pursue further to become an expert on, and set yourself up to leverage it as it becomes more mainstream? The winners in the convergence are going to be those people who spot the opportunities and run with them.

CHAPTER FOUR

WHO OWNS WHAT?

In this third wave, we will take a look at how ownership and finance structures are evolving – from the different ownership structures that are already being implemented to what it will look like in the next 10 to 20 years.This one is really exciting because I think this is probably going to be one of the biggest changes and the biggest disruptors of what we are going to see in the immediate to foreseeable future. Most of this is due to a new technology has just come out called Blockchain and I think out of everything that we are covering in this book, blockchain is going to have the biggest effect on real estate and I will get into more reasons on that soon. But first, I want to share a little bit about what blockchain is because many people get confused between blockchain and cryptocurrency or Bitcoin. Blockchain is simply the technology that cryptocurrency is built on. Kind of like the difference between cash and paper.

I think every real estate investor, every property professional, everyone with even a slight interest in property should be learning more about blockchain. Go learn about blockchain because it is going to have such massive implications for those of us in the property industry. From removing the need for a real estate agent, possibly even killing off conveyancing lawyers, from how we actually store property titles and even how we buy property, it is going to change everything.

If you are a middleman or a service provider of some sorts within the real estate industry, then you really need to be paying close attention and checking out what blockchain is because you want to position yourself to utilise it within your business, before it disrupts you.

You need to disrupt yourself before you get disrupted, so pay attention. Blockchain is going to change the real estate industry forever.

The Sharing Economy

One of the trends that is already well ingrained in today's society is the sharing economy. This is the concept that you do not need to go out and buy every single little thing that you need, because there is somebody else out there who already owns it and probably is not using it at the same time. We have seen this from everything from sharing cars, to tools, to clothing as well. There are many items that have been implemented into the sharing economy. What is really interesting is the statistics that state that the sharing economy is still so young. Projections are that it will increase from a $14 billion industry, which it was in 2014, all the way up to $335 billion by 2025. That is a lot of room for someone to come in, play, and say: "You know what? I'm going to take my own little piece of the sharing economy." You could do this simply by leveraging some of the other waves that we have looked at in this book.

In fact, one possible direction we might see this go in, in the foreseeable future is the convergence of the sharing economy with blockchain in order to create a DAO, a Decentralized Autonomous Organisation, which would then essentially be how communities run. If there is someone in that community who owns a car, and somebody else who owns a tractor, and somebody else who might own some tools, then essentially, it all gets added into this organisation where everybody owns a piece, and everything is shared amongst them. It eliminates the platforms that are already transacting these (think Uber or Airbnb), and basically is going to disrupt the disruptors.

Now, speaking of disruptors and the sharing economy, and considering this is all about the future of real estate, we obviously cannot go much further without speaking about Airbnb, probably one of the biggest players in the sharing economy space, and they became a global dominator with their business model. Airbnb offers the ability to rent out empty homes or other 'useless spaces' within your own home: spare bedrooms, even down to couches simply by monetising an existing asset that people never thought of as an asset. That is probably the best explanation for the sharing economy as a whole: monetising an existing asset that you never actually thought was an asset because it was just sitting around doing nothing.

Airbnb really changed the game. They have been dubbed as "the real estate giant that owns absolutely no real estate." Their business model is perfect. Essentially, they are just an agent for people looking for short to medium-term accommodation. Their story of how they came to be is really interesting as well. They could not pay their rent, but they knew there was a conference coming up in town. Hotels were really expensive, and so all they did was go out, buy an airbed, put it on the floor, and they were able to sell to someone coming to that conference who did not want to pay hundreds for a hotel. So instead, they opted just to sleep on the airbed, and then the next day, they gave them breakfast. Literally "airbed and breakfast." So, that philosophy took off, and they realised there are other people who would have spare rooms, or maybe a space in the lounge, that someone would actually pay to accommodate because it was cheaper than staying in a hotel. Airbnb really came in and stirred the pot, they disrupted the hotel industry, they disrupted the short-term accommodation industry, but at the same time, they actually created a platform for hotels to be able to operate on as well, to get found and receive new guests.

So while they were in there disrupting them, they also provided an actual benefit to them as well.

Airbnb has then allowed thousands, tens of thousands, hundreds of thousands of people to be able to put their properties up and list them, and again, to create income from an asset that people never thought of as an asset, just your spare bedroom. What's more, it has seen a rise of entrepreneurs coming in, and now managing it for other people in what is called an Airbnb host manager.

CASE STUDY

Leveraging the Airbnb platform

Two entrepreneurs I know that have done this really well in Auckland, New Zealand, are Natalie and Brendan. When I first met them, they had some properties in a very good location, but the rent wasn't quite there, meaning it was negatively geared. It was costing them around about $30,000 every year, just in interest payments for their mortgage and other expenses that the rent didn't cover.

Working together, we were able to quickly turn that around to the point that Natalie and Brendan were able to quit their jobs 11 months later. The way they did this was by finding a niche, which was high-profile corporate clients who were looking to stay near the CBD, they were generally staying for a longer period, like that sort of week-long space to two weeks, and in fact, they've had some really long bookings as well. But due to their location and their niche, they had high-quality clients coming in who would pay a premium because they wanted something that felt a lot more like home than simply going and staying in a hotel. They have the majority of their properties in a residential area but they narrowed down where most of their properties were to one luxury development, meaning guests still have access to all the niceties such as the pool, the gym, and all they did was turn their existing assets into an Airbnb rental.

They were so successful that they went out and took over the marketing and managing of other people's houses, as well. For one, they took a lease on another property, and they discovered the difference between what they're paying on the lease and what they could earn through renting it out on Airbnb was sufficient to make it worthwhile for them to do it. They started managing other people's properties who were interested in getting that return, and basically, they acted as an agent, managing everything from the marketing, to ensuring the place was up to spec, running the cleaning crews, managing the linen hire, that side of things, and they took a percentage of the rent to do that. Very quickly, they gained a good little portfolio of Airbnb rentals, and ultimately, like I said, Natalie and Brendan quit their jobs 11 months later. There are many different aspects that you don't even think about when some of these technologies and some of these trends come into play.

Co-Ownership

One element of the sharing economy that we have seen start to come into play is the co-ownership model, especially in the property industry: the co-ownership of properties. So far it has been more common in a closed community aspect or for holiday properties. There are a couple of different ways that you can co-own properties, the first being fractional ownership and the second being crowdfunding. Crowdfunding particularly is more for investment. Fractional ownership is probably more for lifestyle.

Starting with fractional ownership, this is where you have a percentage of the property or maybe even an entire portfolio. For example, you like going to a particular holiday destination. You have friends who like going there too. Instead of buying into a timeshare and not actually owning any of the assets, the five of you come together and pool your cash to own this property outright. You decide amongst you who owns what percentage, depending on how much cash they put in and what the actual entity would look like, whether it is a company or a trust and then you have some sort of time basis to it as well. Fractional ownership is really great when it comes to lifestyle properties, and I will give you a bit more insight into a new business I created using the Fractional Ownership model so that myself and other entrepreneurs could own luxury properties around the world. It is called the Global Lifestyle Club.

Global Lifestyle Club

After the learning experience I had with my coliving space, I pivoted a different direction and started up the Global Lifestyle Club. Interestingly enough, this was more in line with what I had been trying to do in the first place, I just did not know it yet. See, what I was trying to achieve was a place for entrepreneurs who travelled the world to call home - no matter where they were. Every property would be of a similar standard so an entrepreneur could visit a property in Singapore or Berlin and get the same sort of reception, have the same base amenities and know that they would have the key resources they needed close by.

I was celebrating my failure when I met Chandresh Pala, a middle-aged entrepreneur from England who shared with me that his son was about to leave to university. This left just him and his wife in a big house in London, which was not necessary. He told me he wanted to travel the world and was even thinking about maybe buying a second home in Spain, but the problem was, he would always have to go to Spain. After a bit of a discussion where I shared with him what my original plans had been for coliving, we realised there was an alignment and with a change of business model and pivot of clientele we were in business. We soon discovered there were many other high net worth entrepreneurs who loved the idea of co-owning luxury properties around the world that they could use as their home for a week or a month before moving on to another country. As a result we brought on Marc Pinto as another business partner and we fiercely worked out all the finer details of the company before doing a soft launch amongst our networks in early 2018.

What is important to note is that this business is set to take advantage of many of the exponential principles outlined in this book, most notably the following:

> ***Crowdsourcing:*** *we will be establishing a system for our members to present properties that they wish to buy, which if they meet our criteria will be put to the other members to vote upon. This removes us from the property finding aspect and allows members to feel more a part of the club.*
>
> ***Blockchain:*** *Memberships will be secured on the blockchain giving greater security and ease of liquidity.*
>
> ***Sharing economy:*** *Co-ownership is very much part of the sharing economy, and by contributing to the club, members get access to more properties in different locations than they would if they were to simply buy their own second home somewhere.*

Crowdfunding

Crowdfunding, however, is one of the biggest trends to have taken place in the real estate industry within the last five years and is making a real difference in democratising property ownership. In fact, 2017 was predicted to be one of the biggest years in crowdfunding history, because in the United States, it was when the law changed to allow non-accredited investors to be able to invest in real estate through crowdfunding schemes.

Essentially, crowdfunding is multiple people coming together to purchase a particular property. For example, if it costs $100, then 10 people will each put in $10 and own that property together. The big difference between this and fractional ownership is that in crowdfunding there is going to be a lot more people paying smaller amounts. Fractional ownership is usually through some sort of club, and it is probably going to own multiple properties across it, whereas crowdfunding is the opportunity to invest in one particular deal.

It is similar to a real estate investment trust (REIT), where you invest into a trust and someone else buys an asset, but again, in a REIT, it would go out and buy multiple properties, and if one property does well and one property does not, then that is sort of balanced across the entire portfolio. But, if you knew there was a property you did not think that the REIT should buy, there is nothing you could do about it. With crowdfunding, you just put your money into the particular deal that you're interested in buying and your risk is minimised.

This has had a massive effect on who can invest in property, with crowdfunding companies making a lot of money out of this as well. So far, it has been predominantly commercial property that they have been out buying, generally in that $10 million mark. I know, having researched a little bit and looking at what some of the real estate crowdfunding sites are doing, they have found a sweet spot where if it is less than $10 million many investors can afford to own it by themselves. Then there is the $50 million to $1 billion space, which is where many of the REITs, and the big syndicates, and other big players are operating, but there are not too many people in that sweet spot in the middle.

What is interesting is that by 2025, it is projected to be a $300 billion global market, meaning there is still so much room in the crowdfunding space. And, as I pointed out at the start of this book, if we look at what is happening in Australia, New Zealand, and the Pacific region, there is so much room for growth because nobody is doing it or nobody is really dominating.

There are not too many players in this field, and this leaves room for a massive opportunity.

CASE STUDY

Fundrise

One of the big leaders in the property crowdfunding space is a company called Fundrise. It is an American company, which makes it a little bit more difficult for us non-Americans to invest in, but they have really been innovating around what is actually happening in the digital investment space. They started off with crowdfunding, they moved into eREITS, and they are constantly evolving. They are really cool. I suggest going and checking them out and seeing what they are doing. Is there a way you can replicate or do something similar to what they are doing, or is there a way you can just invest with them? Is there a way you can see what they are doing and take it to another country? Check them out here: ***www.fundrise.com***

Global Investing

One trend which we are really just at the start of is investing globally. Now I discussed globalisation back in location, location, location, but this time let's take a look from an investment point of view on how you can own a global portfolio of properties.

Traditionally, most investors will invest within an hour of their home. This is because they want to be able to drive past a property and go see it, to keep an eye on things. And it also makes it easier when you are actually out looking for properties. Then there are a few investors who know that it is smarter to invest in emerging markets or areas within the country that may not be the best lifestyle properties, but they are really great rental properties because they have a high return on their money. This obviously carries greater risk, but has the potential for greater returns as well.

However, we are starting to get to the point where it is actually quite feasible, easier, and you can get some fantastic returns from investing internationally. This might be because you can leverage the currency conditions and fluctuations or it could be because there are absolutely fantastic deals out there. It also allows you to move into different markets. If you are investing locally, you are always at the mercy of what your market is doing. However, if you say: "I only want to invest in one particular market," maybe it is a down market, maybe it is a boom market, you can simply look around the world to see what is happening in that space.

In fact, many digital nomads are starting to invest internationally, especially as they travel to locations such as Bulgaria where they can buy an apartment for as little as $10,000. It makes things look a lot more attractive compared with the $750,000 average purchase price for an apartment in New York City.

Personally, I am really intrigued by what is happening in the global real estate market. Especially as I am watching inefficient markets being gobbled up by hungry investors. As a result, I have started playing in the big data area and have a company creating Artificial Intelligence that can read market signs. I have set up a trial syndicate to see how things work and our investors are getting great returns. The next step will be to create a fund to leverage international markets and make people a lot more money, increasing their cash flow and giving them a better chance for freedom. This is something I am really excited by, and it is really paving the way for that whole globalisation side of things. There are not too many companies operating on a global scale, so there are plenty of opportunities there for us property entrepreneurs to find a niche and become a major player very quickly.

One of the biggest issues with globalisation investing is that it can be very difficult to get mortgages and to leverage your funds. Often, you either have to pay cash, or you have to buy as part of a crowdfunding site. As far as I am aware, there is only one crowdfunding site that allows you to buy across multiple countries: Wealth Migrate. Wealth Migrate was set up by a friend of mine, Scott Picken, and it allows people to invest in different opportunities in different markets around the world via their crowdfunding platform. At the moment, they are predominantly across the United States, the United Kingdom, and Australia, but they are opening up in different countries. They started with a niche for people who are in countries where their currency is devaluing. For example, South Africa at the moment, where the Rand is just continuing to devalue, but also countries like China where people have made a lot of money, and the government tries to keep all that money locked in China. So people are trying desperately to get it out. Investing internationally allows them to get an asset that's going to hold its value while getting their money out of that country and away from a potentially not-so-trustworthy government or into a more stable currency.

Investing internationally can be amazing if you are able to get into emerging markets. There are some countries which are experiencing crazy growth, and if you think about China and the growth it has had over the last 20 years, if only you could have gotten onto the property ladder back then. It has been the same here in Bali, where 10 years ago you could buy 100 square metres of land for $600. Now, that same piece of land could cost you up to $60,000, such massive growth in only 10 years.

Emerging markets are obviously quite risky. Usually they are going to have pretty dodgy governments and often there are many laws about foreigners buying land, but if you can figure out how it works, if you can do it with money that you know you might lose and are okay with that, then the rewards are definitely there.

Globalisation is really something to keep an eye on. If you have been thinking locally, can you think global? Do you have another passport or some sort of other citizenship or residency that you could leverage to go out and get certain benefits? For example, especially when we are talking about Australia and New Zealand, many people have access to an English parent or grandparent that can help them leverage into England. A lot of people have European heritage as well, meaning you might be able to actually leverage it to get some sort of funding within the European Union as well, which is quite cool.

The Death of the Middleman

Moving into the foreseeable future, we are going to see a massive impact on real estate agents, the jobs they do, and how they actually work, because of the massive trend with the death of the middleman. Platforms are emerging, and it's no longer necessary to go to an agent.If you are a real estate agent, you need to reconsider what exactly your job is and where is the value that you provide, because somebody is going to be able to come out and do it much better than you, without you.

We already have artificial intelligence that can write articles. In fact, a lot of news articles you read are quite possibly already written by artificial intelligence and you never even noticed. There is nothing stopping a company setting up the same sort of thing within the real estate industry where you have artificial intelligence writing the real estate listings.

We are also very much moving towards virtual reality where people will not even need to go and visit the house anymore, or at least not initially.They will be able to walk through the house in virtual reality, in much the same way as you can look at photos of the property on the internet at the moment. Meaning, while the real estate agent's job will change, a lot of work will be done prior to the customer actually getting in touch with a real estate agent.You are probably going to get in touch with an agent once you've done a lot more due diligence, and once you've done a virtual walk-through of the property, and therefore the leads that the agent works with are likely to be smaller, but higher quality.

But, again, due to blockchain and smart contracts, it really is going to be interesting to see what direction the real estate agent's industry takes. Smart contracts are built on blockchain, which essentially allows for autonomous transactions to happen. It says that when this happens, this gets triggered automatically. And, because it is a smart contract and it is built on blockchain, it cannot be broken. It is almost like a promise that cannot be broken. This is very much the role that a lawyer plays today. They take a legal undertaking or a legal promise that if you do something then they will do something. For example, if a purchaser supplies the money then the lawyers have an undertaking that once the money has been provided then the title will change hands.

Taking a look at real estate agencies, it is going to be very interesting to see which direction they do go. One company which is leading the way in disrupting the real estate industry is Matterport. Matterport is a company that allows for digital walk-throughs of any property. They have a special 360-degree camera that will photograph the inside of the home and stitch every photo together to create a seamless representation of that property. This allows you to then walk through in virtual reality – you can do it on your computer screen if you do not have VR as well.

In the future, I imagine it will go something like this. You will have your virtual reality goggles on, you will be walking through, if you are interested a virtual real estate agent will pop-up that you can have a real-life conversation with. This agent will be powered by artificial intelligence. It might be an avatar of the actual listing agent, it might be a generic avatar. But you can sit there asking questions, and ultimately this Artificial Intelligence will probably do most of the work of the agent in closing the sale. If we then have smart contracts, that allows the purchaser to sign an online contract, which could be conditional, but essentially removing the need visit a property altogether.

Because of blockchain we will have all the files relating to that property within easy access, so that you can do all your due diligence there online at the time. You can simply ask the agent, "Hey, can I see the building report?" And that could be sent through. The same can be said about council file requests. All paperwork could appear in front of you. So by the time you are signing that contract, you are pretty much stating either, "I want to view the property in real life", or, "It's subject to finance."

Again, with blockchain, you can already have all your details listed on the blockchain and we are heading towards a time where we're going to have digital IDs secured on blockchain so all that information can already be there. The same app could approach the bank via blockchain, and the AI behind the bank can almost instantly approve you. So, the transactional speed of purchasing properties is going to happen a lot quicker.

CASE STUDY

Rev 3

One New Zealand company that is really playing in that virtual reality space is run by a friend of mine, Pleayo from Christchurch. His company – Rev 3 – is working with virtual reality to provide developers with the ability to allow potential customers to walk through the property and see what it will look like once it's been built. At the moment, you are probably familiar with buying off the plan. This usually involves skimming through a folder where you look through all the plans and artist impressions. Now, you can actually put on the virtual reality goggles and do a walk through of what your future property will look like.

This is the bedroom. This is what the kitchen will look like. This is the lounge. Here's the deck. And then, you can also change things, so if you want something a different colour, or if you want different flooring, you can simply select the menu on your virtual reality, change it and see what that would look like. So that's really cool.

He is also playing with augmented reality, which I found really fascinating. For one deal I was looking at utilising VR, but unfortunately, we could not come to an agreement. But, it was a large tract of land and if we were using augmented reality then we could have actually had it where we can go out to the land and with our perspective buyers, put on a Microsoft Hololens and they can see the structures of how everything would look on the actual land. We can then shrink it down to a specific size so they can see what it would look like as a whole and then blow it back up again to a 1:1 scale where they are now walking through a digital representation of what would be built on that land. Even better, they can still go into the house once it's built and do exactly the same walk around, having a look. So there is a little bit of a difference between augmented and virtual reality, but ultimately you are able to leverage both within what you're doing and Rev 3 is doing cool things in that space.

As I stated before, blockchain is going to eliminate most of the middlemen within the real estate industry. Looking at the longer term, that 10-year mark, we are going to see a lot of this start to change. We have already talked about real estate agents but it is also going to affect lawyers, conveyancing lawyers in particular, because blockchain eliminates that need for the undertaking, that legal promise.

If you are a lawyer and you specialise in property, you already need to be thinking about, again, what is the actual value you provide because, at the moment, you are very transaction-based. Anybody within the transaction-based industry is going to be eliminated. What you need to do is look at what's happened within the travel industry. Back in the day, 10-15 years ago, a travel agent's job was to book flights for people or to book holidays. You can do all of that online now. The travel agents that have survived are the experiential ones, the ones which provide an experience for their guests. How can you find a way to pivot what you are doing and create more of an experiential business model? Because, if you're just relying on doing transactions, the need for you is going to disappear.

We are going to see the same with many other middlemen such as title registries. They're going to disappear in the sense that a title can simply be secured on the blockchain as a digital token. If you had that happen, we don't need to be paying to access property data anymore. We do not need to be paying to transact the title because it is going to be like one cent. It is going to be ridiculously cheap. All of these sorts of things are going to change simply because blockchain can do it all so much better. If you are some sort of middleman, you need to look at how you can move from a transaction-based business into an experiential business.

End of the Physical Property Title

Also in the mid-term future, we are going to see the removal of a physical property deed because, again, this will just be a digital token secured on the blockchain. In the same way that you can own a bitcoin today, you could own the deed to your property. Countries like New Zealand already have something similar, where you do not receive a physical title. It is just a copy. The title is secured against your name in a government registry, and there is nothing stopping that from simply evolving into being a digital token on the blockchain. Some countries are already starting to experiment with this, with Estonia being one. Dubai is obviously one of the biggest leaders in this industry, and in innovation and technology, so they are definitely having a look at what they can do in that space.

If we remove property titles and put them on the blockchain, it is going to allow for much quicker transactions. It is going to allow for easy due diligence, because not only will we have access to the title, but you will also have access to all files and all government council files related to that property. They will no doubt be somehow attached to the title.

It is interesting because this will bring with it new issues that we will need to overcome. We have already seen in the cryptocurrency space, people who forgot their passwords/private keys losing all their cryptocurrency, so the question is: if you lost your password, could you lose your house as well? These are a few things that whoever is designing the system needs to think about. It could possibly be something similar to what New Zealand already has with an adaptation to it that you have the government central registry where the actual, physical title is held in trust for you, with you receiving some sort of token, which is representative of the title but not the physical title? Meaning, if you did lose your password, the government still knows it is you. Or would it simply be like bitcoin, like a cryptocurrency, where if you lost your password, you literally lose your house? I do not know. We will probably see a few different versions as countries around the world start experimenting. Also, as different models evolve, we'll see some different ways playing out.

A Stock Market for Real Estate

A little further into the future, however, we are essentially going to see the real estate market turn into something similar to the stock market where you will be able to buy pieces of properties. Obviously, you can already do that via crowdfunding at the moment, but it's not liquid. I have already met people who are trying to do this so no doubt there are others trying to do this too.

The biggest thing about a stock market for real estate is the way it is going to change how we invest, how we are saving, and how we fund properties. For example, if you are thinking about buying your first home, traditionally you save money in the bank. However, it would be possible for you to go find the home you want to buy and then simply have a 10-year goal where you start by renting the property and the money you are saving, you are actually buying shares in the property, decreasing the amount of rent you have to pay. Other investors could own the other shares in the property, thereby receiving a percentage of the rent that you pay and possibly any capital growth from their shares as the market moves.

Once we have a stock market for real estate, it is really going to beg the question of what is the difference between possession and ownership. So, you could raise funds by finding 10 other people who want to invest in that property. Instead of having to get a mortgage, they would invest and you would pay them out a percentage, whether it is for rent or interest or whatever it is called in the future. So, for example, if you need to raise $100,000, you might put in $10,000, raise $90,000 through other investors and then you take possession of the property, which gives you the right to live in the property. The other nine people all have an ownership stake in the property, but they do not have the right to possession. It is kind of like a license to occupy.

So, it is going to be very interesting. That is a lot further away but it will be very fun to see how exactly that plays out. I imagine we are going to see things start popping up probably in the next five to 10 years, where private companies are experimenting with their own portfolios, allowing people to have some sort of share across their whole range of properties. It will most likely be a type of crowdfunding company that pivots into this model.

As I said at the beginning of this chapter, ownership structures and financing structures are all on the verge of being disrupted. Blockchain is going to be a massive cause for this disruption, but there are things you can already be doing, and things you should definitely be aiming at. I think this is probably the most important module in the entire book, where if you can figure out how to utilise the blockchain, how to figure out some of these different financing structures, then you pretty much have access to all the capital you would ever need. You just need to leverage. There are people wanting to invest, who can invest, so you can simply need to come up with a model for them to be able to invest with you. There is so much opportunity in this space. The Fintech area is just exploding. There are new emerging companies every day doing amazing things and it would be awesome to see someone like you doing something like that within the real estate sector, as well. There is so much opportunity for people within this financial space, within this ownership space, and it simply comes down to 'who owns what.'

CHAPTER FIVE

LIVING ON PURPOSE

This chapter is all about how people are now deciding they want to live somewhere for a particular purpose. In essence, their environment is starting to reflect their purpose more and more. Rather than simply living where we grew up, we can actually choose to live anywhere in the world. We can choose to live with pretty much whoever we want, based around common beliefs, rather than just a need to live somewhere. We are starting to see this happen around the world where everything is becoming a lot more purpose-based.

Community

One of the biggest trends already in place is community. It is all about community. There is a trend now of entrepreneurs who all want to live together, which is exactly what Silicon Valley is. Then, we are also starting to see areas where people who are more environmentally conscious want to live together, and where people who want to be healthy and more spiritual want to live together, so there is this massive trend towards that. We see it in the student space where students all want to live together and we're even starting to see it in the professional millenial age bracket who also prefer to live together rather than on their own.

One company has come up with a concept of mixing community with environment and technology. These are ReGen Villages – **www.regenvillages.com**. ReGen is a Scandinavian company, but they are starting off their first test site in the Netherlands, and it is a chance to build a community from the ground up that is modern and utilises many different technologies for it all to be self-sufficient. It bases its whole philosophy of families coming together to live a more natural lifestyle. If you take a look in every country around the world, there is going to be some sort of similar community or some aspect that you could either go and join to be with the people that you want to be living with. Maybe, if you are a property developer, you could tie in a similar community aspect. People will always seek a sense of belonging and a community is fantastic way to fill that need.

The best way you can utilise this trend is by building more community around what you are doing. If you are a property investor, is there a way you can create a community amongst your tenants, or amongst your partners? Luxome – **www.luxome.co.uk** – is a Manchester-based company that has done this really well with student accommodation in England, and they did it with luxury, high-end student accommodation. Luxome was able to build the community and a brand and this allowed them to be the industry leader in this market.

If you were to build an online community around your property investments, this would allow you to have better connections with your tenants and you would know what is going on. It would also allow them to connect with each other, which is probably a good thing unless you are a really bad landlord, and it provides an extra type of value that people do not already have, and that other people are not providing either.

Obviously, community is going to be more important when we are talking about rent-by-the-rooms and houses of multiple occupation. This has absolutely exploded in the States, where companies have popped up such as Common

—**www.common.com**—and they are describing themselves as boarding houses for professionals - professionals who have come out of university, and are now working proper jobs, but still love that community aspect of when they were at university. This company has 20 to 40 bedrooms within a property and you get to live with your friends. You get to live with like-minded, similar people. You have always got something to do. There is always a community aspect, whether it is upskilling professionally, upskilling personally, or business development.

There is nothing stopping you doing the same with a boarding house in your industry. In fact, boarding house owners are sitting on an underutilised gold mine. If you own a boarding house, you could simply add a sense of community to it, and if you do not own a boarding house, you could partner up with a boarding house investor and build the community aspect to it, pocketing the difference in rent that you achieve. At the moment, boarding houses, particularly in New Zealand are kind of frowned upon and looked down upon. They are somewhere you go when you have nowhere else, yet, if you look at the States, it is where there is a list of thousands of people who want to get into them. So, just that mindset shift would really allow someone to come through and build a massive community, to build a massive company out of this very quickly, and the best part is you don't even need to own the asset.

The other direction that community is taking is within existing communities themselves where they are becoming tighter. There are a lot more community initiatives going on which are being created to bring the community closer together, usually around gardening or schooling. Communities are saying: "Okay, we want to now be known for something or we want to implement a certain type of lifestyle within what we're doing."

One way to leverage this into your existing business would be to simply start something up. Get together some of the key influencers within the community and start something. This then creates a financial value for you because a tight community means people are going to want to get into that community. It is going to push house prices up and your assets are going to go up in value as well.

Can you start something up within your existing community? Or, can you just go out and create a community that you want to live in? That is exactly what I have done here in Bali. I took on a seven-bedroom villa so I could have other entrepreneurs coming and living in my little community. It means when I wake up, I am having breakfast with an entrepreneur. When I come home from work, I am sitting here having a beer with an entrepreneur. It means we are constantly having a higher level of conversation.

Housing as a Service

One trend we are very much right at the beginning of is housing as a service. This is very much geared towards people who don't want long-term leases and want to rent a fully-furnished property. All they are looking to do is utilise your accommodation on a medium to even potentially long-term basis. We saw this with the rise of Airbnb, where people decided to actually start living in Airbnb properties and then that is being leveraged across into coliving. We are seeing people who are coming in and they do not want to own their own furniture. All they want to do is be able to come in, sign a contract, spend a certain amount of time in a property knowing they do not have to worry about sheets, they do not have to worry about cutlery, they do not have to worry about paying utility bills. They simply utilise a house as a service. We are going to see this become a lot more common as we all become more global, travel more and prefer not to have a single physical home. Obviously, some of the population will, there is going to be a lot of people who do not want a physical home that they live in for the next 50 years.

That is one aspect that you can get into straight away, whether it is leveraging the Airbnb platform, or whether it is finding some little niche of people who are going to be out there looking for housing as a service.

Coliving

Another trend which seems to be gaining ground, but is still very much at the start, is the coliving movement. Coliving spaces are very much in line with what we have been talking about with community and housing as a service, but is now talking about generally more medium-term accommodation. Coliving, at the moment, is targeted towards travellers who want to spend a medium amount of time in a particular country, staying at a particular property. Coliving is what a lot of the digital nomads are looking at when they are going somewhere for maybe two or three months.

Coliving is right at the start of a wave that we have seen the hostel and backpacking community go through. We saw big brands emerge, but we also saw the independents emerge and now we are seeing exactly the same in coliving where you have got a few big brands such as WeLive who is more targeted towards the professional market and Common which is also targeting domestic customers.

There is yet to be a clearly defined definition of coliving. It is a split between domestic, which I would call people of the same country coliving within their space within that area, and then foreign which would be the travellers who are coming and coliving together.

When we are talking about the travel side, it is generally for medium-term accommodation. For the traveller, coliving is a very attractive service because it means they can arrive in a new country and instantly be tapped into a network. All the services are provided for them. They just need to show up with their bags. We are seeing lots of independent coliving spaces pop up around the world and we are yet to see one true leader take the stage or take on multiple countries. But, that just means there is a massive opportunity there. In the same way that we saw this absolutely huge growth of co-working spaces, we are going to see the same in coliving. A lot of coliving spaces are attached in some way or another to a co-working space – whether it is the co-working space adding on coliving, or whether it is coliving opening up a space that's available to the general public to use.

There are a few aspects of what I am going to talk about within the foreign traveller type coliving accommodation that is a necessity. It is community. It is good WiFi and then a niche product or area. I actually had a go with coliving myself when I first took on this property here in Bali. I took it on with the vision of having coliving. I wanted to set up coliving spaces all around the world. However, I found out pretty quickly that it was not quite worth it for me, mostly because I wanted to get high-quality entrepreneurs and then I discovered the type of person I was going for did not really want to co-live. They would rather be part of the community but not quite in the same house, not under the one roof.

The other aspect that did not make it viable, was that I could not get the price point I wanted without having that big recognisable brand already. The minute I started competing with other homestays it did not make it financially viable. I turned what I have here in Bali basically into my home and then shared it with other entrepreneurs as just a houseshare-type scenario. But, there are people out there doing it well. The first thing I would say is go check out **www.coliving.com**. That was started up by a friend of mine Daniel Beck and it is the main platform at the forefront of coliving, where he is connecting with many other spaces around the world. If you are interested in opening a coliving space, you will find many different ones from around the world on the site and you can research how they work and the prices they charge.

The one person I know who has done the independent side really well is a girl called Maria and she runs ReStation (**www.restation.co**) over in the Canary Islands in Spain. She did it simply by leveraging other people's assets and by growing as she needed to. Instead of simply taking on a lease for a really big property, she started off with a smaller one and the coworking space and she bought on more properties as she grew. While it is not just one property, it's a community aspect within walking distance from each other.

Work-Life Balance

Ultimately, one of the biggest things driving this whole living on purpose side of things, is a work-life balance. We are seeing it across all sectors.

It is really prominent in the millennial sector, when you are finished uni, you have done a few years of work, and then you are like: "Actually, I just want to get out and travel more". That 25 to 35 age range, it is really strong there and people do not want to do the 9-to-5 lifestyle. They want a much better work to life balance. They want to be able to make their own money, hence the massive rise in freelancers, contractors, and remote workers. As people discover that they can live and work from anywhere, they are deciding to go out and do just that.

However, it is not just within the millennial age bracket. We are seeing many CEOs simply say: "I do not actually need to be in the office 24-7 anymore. I am going to live in a better part of town, maybe a further commute, but not having to come into the office so much." I have even met CEOs and prominent entrepreneurs, founders, who have no fixed abode. They are just constantly travelling. They have set up a remote team, a remote lifestyle, and they are able to do whatever they want, and fly, and just go live in whatever country suits their purpose at that time.

When you are in business, the energy of it is constantly changing. We call this seasons, and CEOs can channel this energy, finding the country that allows them to best tap into that energy. Let's say, if you are looking for more grounded energy, the CEO might go to a country that has that energy - the same reason I moved to Bali. If you are looking for a more upbeat, energetic sort of energy, with lots of connection, and brainstorming fresh ideas, then they might just uproot themselves and go live somewhere like London, New York or Berlin for three to six months.

We are seeing companies like Virgin who say: "You do not need to work the 9-to-5 anymore. As long as you are getting results, you can take as many holidays as you want. You can live and work from wherever you want."

We are also seeing it within the retiree sector. Australia is famous for having grey nomads, which is when you simply sell off everything you own, jump in a campervan or four-wheel drive towing the caravan, and just non-stop drive around Australia. No doubt, that is probably prominent in a few other countries as well.

This whole living on purpose thing, it is not just any one particular type of person. It ranges across everyone. There is definite room for whatever niche you can find within that industry.

Living Cities

Looking a little bit further afield at the foreseeable future, we are starting to find living cities. This is a direction that many cities have said they want to go in, and they are starting to make those changes now. We can start to expect a lot more living cities coming into play. Singapore is already playing with this a little bit, and you only need to walk through the CBD to find buildings completely covered in green, covered in plant life, which is attracting the bird life, which then creates that ecosystem of growth.

One city that is dead-keen to get this happening is Beijing – known as one of the biggest polluters in the world and absolutely suffocating its citizens with smog. There are many concepts coming out now of how Beijing can turn into a green living city. From massive forest towers to all sorts of new devices which are actually going to suck the smog out of the air and turn it into coal. In fact, you can even buy some jewellery which has taken the smog from the air, compressed it down until it becomes a gemstone, and then it is turned into a ring, which I think is pretty cool.

There has been a rise in the number of rooftop gardens across New York, including turning old train lines into parks. Going from a concrete jungle to more of a hybrid where it is possible to have a lot more park life. We are going to see a lot more of this in conjunction with what I was talking about with self-driving cars, as roads become less needed, they can be shrunk down in size and the sides of the roads will then be turned back into parks. We are going to see many more cycleways and other personal mobility vehicles implemented. Really, it is going to create a city that is in tune with nature, where it is in tune with pedestrian, cyclists, and everyone can flourish and live together. We will see a lot of animal and birdlife come back into the cities, and it is basically almost going back to our roots.

On the residential side, one company that is doing this whole living city side really well is Green Magic Homes – **www.greenmagichomes.com**. They have created a sort of dome-like house structure where you can pack earth on top. It is almost like a precast of an Earthship. You can buy it, stack the earth on top, grow your grass, grow your fruit and veggies, all these types of things on top of this magic home. I think that is really cool. They had so much demand when they got started that they have opened up factories all around the world. They are growing so fast. And I think these sort of Green Magic Homes are going to start popping up in cities very soon. At the moment, they are more rural because they fit in with that landscape so well.

Seasteading

Looking into the more distant future, it is going to be possible for entrepreneurs and property investors to not only invest in property, but also to create a whole new city of your own. In fact, not just a city, but your own country. This is something called seasteading, where you build a floating city, and as long as it is outside the territorial waters of an existing country, you can essentially claim to be your own country and to make all your own rules.

Now, while this sounds a little bit far-fetched, it is actually starting today. There is an organisation called the Seasteading Institute – **www.seasteading.org** – and they have signed an agreement with French Polynesia to go out and build their own floating island, their own floating city within the territorial waters of Tahiti. This is going to be absolutely massive.

At the moment, it is backed by a lot of Silicon Valley tech entrepreneurs including Paypal founder Peter Thiel – partly because it means they can create their own tax rules, but it essentially gives anyone the ability to create their own country or their own city in conjunction with community and be able to invite anybody to be citizens. This means you can build it around whatever purpose or theme you like, and you can create your own rules. This is supposed to be changing the world, simply because it is going to allow us a lot of opportunities to experiment with different types of governance. There are going to be countries that pop up where maybe the legalities around healthcare are a little bit lenient. So, they can experiment with healthcare regulations which currently are illegal. We are going to see countries pop up which have no religious beliefs and so that will change the culture of the country. We might see things such as global passports, open borders, and cross-experimenting all appearing because of the possibilities from thousands of countries experimenting in their own way of governance.

Seasteading Institute is leading the way in this area. The first project is supposed to begin in 2 years time and the team at Seasteading institute hypothesise that in 30 years time there could be thousands of floating cities – I call dibs on calling mine 'Atlantis.'

CASE STUDY

City of Mériens

The City of Mériens is a concept envisioned by French architect Jacques Rougerie as a tribute to his love of the sea. The giant floating city is shaped like a manta ray and is designed to conduct research on the surrounding ocean. As a result, it is able to house over 7,000 researchers, professors, and students and will be equipped with laboratories and classrooms. This is such an exciting project because it is right at the beginning of the seasteading wave. The city would be completely self-sustainable and produce no waste, gaining its energy from wave movement, wind, and solar. While the City of Mériens is only a concept, it is not a far-fetched idea given that in 2014 Jacques raised $430,000 in crowdfunding to build the Sea Orbiter – ***www.seaorbiter.com*** *– a much smaller version of the City of Mériens, which is capable of housing 18-22 people, but is also fully submersible. This is an indication that seasteading and self-sustainable floating cities are definitely going to be part of our future. The question is, when will humanity transcend to the oceans to live?*

The world is connecting in with their purpose, and as they do, this is changing how, where and with whom people choose to live. As property entrepreneurs, this opens up a lot of opportunity to create spaces specifically for these people, be it a coliving space, a whole community (online or offline) or maybe one day even starting up an entire city dedicated to one specific purpose.

So, as we come to a close on this chapter, you must ask yourself: how are you going to incorporate some of these themes into your current business? Are you going to experiment with community, could you leverage a boarding house, or are you going to start planning for the time that you can build your own country?

CHAPTER SIX

THE DEMONETISATION OF HOUSING

It sounds pretty absurd, the idea of houses being completely demonetised, and trust me, I know. Here I am a property entrepreneur – someone who makes all his money from property – not only saying that housing is going to become free, but also wanting it.

First, let me explain my reason for wanting this trend to happen. I believe that if we can get to the point where everyone on earth has shelter and that home can provide them with electricity, water and food, then we raise the standard of living across the globe. We set a new benchmark of how far someone can fall when times get hard and more importantly, if everyone is fed, everyone has shelter, then we are able to rid the world of crimes of desperation. What is more, I believe it will encourage entrepreneurship and for people to take on the challenge of solving many of the world's biggest problems. Think about it. If you knew you could try to solve the biggest issue closest to your heart and if you failed you would still have food on the table and a roof over your head, would you do it? Of course you would. Hence why I believe, in the long run, it will be a major asset to humanity.

I should also point out that this does not mean you can simply go out and get an awesome, amazing mansion for free. What I mean is, the cost of housing in many areas will come down, but also some business models will rise which means people can live in the house for free. I mean, I was provided accommodation and food when I worked on the mines and oil rigs, so the idea is not that far-fetched.

To understand why housing could become demonetised, we need to look at Peter Diamandis' Theory of the '6 Ds of Exponential Technology'. Peter Diamandis is an entrepreneur and a futurist who specialises in building businesses, utilising exponential technology. He has a few businesses under his belt including Planetary Resources, a company which is going out into space to mine asteroids, as well as Human Longevity, Inc., whose mantra is "100 is the new 60." He is playing with a lot of new technologies and pushing the limits, and he has broken down everything that has been happening ever since the advent of the computer and exponential technologies into the 6 Ds,. So, let's take a look at each D, what it means, and we can have a look at what that might mean for the housing industry.

The first D is digitising. This is when things first become digitised. If you think about the digital camera where it took photos and cancelled out the need for it to be printed onto film, simply by digitising the photo. It turned it into 1s and 0s and stored it as data on a chip. Anything that becomes digitised enters the same exponential growth that we see in the computing industry. It becomes easy to access, easy to share and easy to distribute, so, of course, for housing, we'd most likely see this with 3D printing of housing and house plans, etc, becoming widely available.

We may see it with the different virtual worlds, augmented reality with the different worlds that we can have there, and then of course, way in the future where we leave our human bodies and become digitised ourselves – but that is a story for another time.

The second D is for deceptive. This is when it first becomes digitised. However, on the exponential growth curve when it doubles from 0.01 to 0.02 it is so small, that it is almost unnoticeable. In fact when you look at the whole start of the curve, 0.1, 0.2, 0.4, 0.8. 1.6, 3.2 the growth is so small it is indistinguishable from 0.

This is exactly what happened to Kodak, who actually invented the digital camera. However, because the growth was so deceptive, they never actually realised the true power of it until it was too late.

Deceptive is the initial period where we do not see much happening, and so, as soon as it does become available, like 3D printing houses at the moment, it is a new technology. Things have become digitised, but it is still in that deceptive growth phase until we start to see a wide distribution of 3D printers, and until other things happen that influence it so that it would become a lot more mainstream, so that it can happen faster.

There are probably other aspects of the building industry or even the construction process that are already moving into digitalisation, but it is in the deceptive phase so we do not even really know what is going on.

The third D is disruptive. This is where we move from deceptive to disruptive on the exponential curve. Disruptive is very much when businesses start going out of business, when they start going bankrupt, when old business models become redundant, when you really need to change up your business, or die.

I see a lot of this happening within the car and real estate market within the next 10 to 15 years where it really starts to become disruptive. This is where I talked about the convergence and many of these waves and forces are coming together to really disrupt the industry.

Disruptive is when the changes are so big and dramatic that by the time you start trying to play catch up, it is almost too late.

That is why it is so important to look into the future to see what trends and technology are coming, otherwise we get complacent, we get disrupted and left behind.

Then, we come to the fourth D, which is demonetisation. This is how that digital photo suddenly became absolutely free. Nowadays, you can take photos with your phone and it is absolutely free for me to share it on Facebook or Instagram or anything else I want to do with it. In the past, you would have to pay to print it out. You could only take a certain amount of photos because it was all on film, so you would have to buy the film, you would have to pay to get it printed, and then you would usually have to store it in an album. It was not demonetised at all. This is where Kodak went wrong, because they thought they were in the printing and film industry, as soon as the digital camera made film and print redundant, they had no business. Interestingly enough, the same year Kodak went bankrupt was the same year that Facebook bought Instagram for $1 billion.

We are already seeing demonetisation occurring in the 3D printing world where we can go in and get blueprints for free and just download and print whatever we want, and we will definitely see that happen within the housing industry as well. Think about using recycled material to 3D print a house, it will cost you almost nothing.

It is also probably going to happen in the energy sector where things are going to become so cheap. I mean, we already have printable solar panels, and so that industry just needs to move from deceptive to disruptive, and then that will move to demonetised as well, meaning, we can pretty much print solar panels and get solar energy for free.

We are definitely going to see this happening in parts of the real estate industry within the next 20 years. I do not think it will ever make the entire real estate industry valueless, but there will be a certain sector or a certain business model which will mean many people get to live in their houses for free.

Next up, the 5th D is dematerialised, which is where you are actually removing the physical object. If you want to think about smartphones today, they have GPS units, they have radios, they have a "computer" pretty much. They have so many different things that you are able to dematerialise the existing hardware and turn it into software.

If you think about how many apps you have on your phone, back in the day, you would have to have a Nintendo Gameboy or you would have to have buy the individual game, and then you might only own five individual games. Now, you can have a hundred, a thousand different games on your phone. You can download them, or delete them as quickly as you like.

We have already talked about having virtual furniture, virtual artwork on the wall, so that's all dematerialisation as well and, again, there will start to be things for free in that space as well.

Finally number 6, is democratised. This is when it becomes available for pretty much everyone. When it becomes free, it gets more access to people. In the same way that Facebook democratised social media, it democratised people into having the ability to have a say, which then led to things such as the Arab Spring.

We are going to see a lot of this housing technology become democratised to the point where anybody can build any type of house, anybody can have access to land, anybody can almost do anything within the housing industry. To be honest, this is starting to look way into the future now, but one day it will happen.

Those are the 6 Ds that Peter Diamandis talks about. If you are not following him already, you should definitely go search for him. He is absolutely amazing in business. Another person to check out is Ray Kurzweil, an exponential technology futurist who has predicted so many things so accurately as well, so check them out.

Tiny Homes

Talking about the demonetisation of housing, one thing we are already seeing is smaller homes, in fact not just smaller, but tiny. The tiny home movement has spread across the world. Particularly with a few different types of people, e.g:

Millennials who are either like "Hey, we cannot afford to get on the housing ladder", or "Actually we really do not need that much." So they go for that minimalist lifestyle, living in a tiny house because they really do not need that much, or do not want that much. Even some billionaires have decided to live in a tiny house just because they love that minimalistic lifestyle.

Pop-up communities of tiny houses are starting to emerge, as well. All it takes is a park-like setting - this could be co-owned by the community or it might be owned by one person and leased out to the other renters – but it means you can fit more people in a smaller area, and they all own an asset.

So, this is definitely demonetising the cost of housing, because instead of paying $700,000 to buy a piece of land to build a house on, you are paying between $10,000 to $50,000. Iit makes living a lot more affordable. A lot of tiny houses are off the grid as well. We are about to get into self-sustainability, but it means that the cost of living is so much cheaper. When the cost of living decreases, the need to work harder decreases, because you do not need as much money to live, so you have that better work-life balance. That is precisely what we were talking about in the living for purpose chapter, of how more people want to be spending their time doing what they love, not working for the sake of earning money to earn a living.

The tiny house movement is really exciting, particularly in the homeless sector where there are a few organisations out there experimenting with tiny houses. There is one particular one in Detroit that is working with homeless, or less well-off people, to help them become a homeowner, by having a shared land distribution with the ability to 'rent to buy' a tiny house. This then gives them an asset that can appreciate in value, be passed down to family members, or even sold to upgrade to a bigger house.

So, tiny houses are a movement that is already well-established that you can jump on board with and better yet, depending on what the rules of your country are, often you do not even need a license to build. So again, it will demonetise the whole council process, by removing those unnecessary costs.

What is it that you could be doing with tiny houses? Is there a way you could implement it by putting it on an existing piece of land and then renting that out? Could you be putting it on the back of an original house and then renting that out separately, or renting it out together as providing extra room? Or, could you be buying up a whole heap of tiny houses and creating a tiny house community? And, then this ties in with what we are talking about in living on purpose, where you have now got the community aspect, you have got the demonetisation, you have got tiny houses, and that is a reason for people to come together, live together, and turn it into something awesome.

Self-sufficient Homes

Another established trend that can be implemented immediately, is a massive shift from fossil fuels to renewable resources and self-sufficiency. In fact, there is a bit of a war going on with some countries deeming a self-sufficient house to be illegal. I can tell you who is going to win that war: the people. But ultimately, the movement has already started as we are moving towards a more sustainable lifestyle. As a result, in the future, nearly every house is going to produce its own water, create its own electricity, and grow food.

The biggest movement at the moment is in the energy sector where we are heading away from fossil fuels and into more renewable resources, such as solar and wind. For the home application, there are some really exciting new inventions, which are very much in that digitising/deceptive stage. One sub-trend is turning everyday items into generators of electricity. With the most well known recent addition being Elon Musk rolling out the Tesla RoofTile. Interestingly enough, this solar panel tile is not in direct competition with other solar panels, it is in direct competition with roofing tiles. And we are going to see this happen more and more.

Technology exists for windows too, where your normal pane glass window in your house will be able to produce electricity as well. This is done by aiming at the spectrums of light that the human eye cannot see, so infrared and ultraviolet. So it is still a clear, transparent window that you can see out of, but it only targets those two extreme ends and creates electricity from that. In Portland, Oregon, a startup company is putting generator turbines in water pipes, so that as water flows down the pipe, as it's always going to do anyway, it turns a turbine, which then creates electricity. No doubt in time we're going to see this on the home level as well, where the water pipes coming into the home will also have little turbines to create electricity as the water gets piped in.

One of the biggest issues we have been facing in recent times has not so much been solar energy collection, but rather energy storage from that solar. Obviously, Tesla is also leading the way with their advent of the Power Wall. Whilst this was absolutely amazing and a fantastic innovation, the real value comes now that competitors have jumped in. There are now other companies out there competing with the Tesla Powerwall, who are creating similar style home batteries, and are initiating price wars. Solar energy has already become the cheapest source of electricity in the world, and we are also seeing things like China building the world's biggest floating solar farm in order to start creating renewable, clean energy. And the funniest thing about this is that China is known as one of the biggest polluters in the world. They actually created this solar farm on a flooded coal mine. So, they are very much showing us that if they can do it, anyone can do it, and they are definitely leading the way in that industry.

Not only that, there is also a rise in innovation to create electricity from less alternative sources. We have always been able to do it from hydro, solar, wind etc but Dutch scientists have discovered that they can actually create electricity from plants, from the soil. Talk about going green! Interestingly enough, the electricity is created by the plant discreting unused minerals, so I guess you could call it powered by plant poop. Plants aren't the only poop we can harvest electricity from, we have also discovered that we can create it from our sewage waste as well, so expect the toilet to get a major technical upgrade in the coming decades as well. Overall, we can expect to see a lot more sustainability with the house as a closed ecosystem, where the house is starting to turn all the waste products back into electricity to be consumed within the household.

Plant-based Meat and Vertical Farming

Looking into the foreseeable future, one of the interesting trends which has a technology component to it as well, is the move towards plant-based meat and vertical farming. This is important because it will remove the need for farmland, which is, in turn, going to create an absolute abundance of land. Looking at vertical farming first, vertical farming is the idea of stacking plant upon plant upon plant, and getting much more produce per acreage than a traditional farm. It is indoors and it is run by scientists, not farmers. It is really done to be able to get as much efficiency as possible. It uses hydroponics or aeroponics, and sometimes even aquaponics, in order to grow the food. It is best for leafy greens and light vegetables and fruits.

We are not going to be completely done with farms yet, but it is definitely going to dictate what we can grow at home. Products already exist, where you can create your own mini vertical farm in your kitchen. Supermarkets are starting to implement these so that they can grow fresh produce right on site. In fact, if we start looking 20 years in the future, I imagine we're going to have food produced a lot closer to where we consume it. If you think about a city centre supermarket, it is usually located at the bottom of the building, and then maybe a few other shops, and an office block on top. We might see a complete change, where you will now have the supermarket on the bottom, but instead, it will be a farm on the top. There might be the herbs growing on level one, lettuces on level two, and then radishes growing on the next level. They will be picked and packaged on-site and taken down to the supermarket, or maybe even shoppers will be able to pick their own produce direct from the farm.

Vertical farming is going to change what we consider is a traditional city. Traditionally you had the CBD with all the tower blocks, and you had the city ring area, which is probably more apartments, units and townhouses. Out to the suburbs, you have got more single dwellings, .then it turns more rural, and then you have got farmland. If farmland is no longer relevant, because the food has been growing in the city centre, then what will this mean for the farms?

It is really going to change what we consider the food bowls of the cities. The question is, is it going to change what food is growing in those food bowls? Is it going to change the need for the food bowl altogether? What does that mean? You look, and there are a lot of farms that are really close to cities, that are so valuable, because they are large tracts of land close to the city. But, if it becomes no longer valuable simply because you can grow much more food in less space, more efficiently, feed more people, and bring the price of food down without having to have transportation costs, what is that going to do?

The second part is plant-based meat. You can argue that at the moment, with vertical farming, we are still going to need farms for raising sheep and beef. But, there is a massive trend towards plant-based meat. I am not necessarily talking about veganism. I am talking about the new technology that has come out, that is reproducing meat from a cellular level completely out of plant proteins. Meaning, it tastes like meat, it bleeds like meat, and in fact, you probably would not even know that it was not meat if you had not been told.

If we are moving towards this side of things, there is a very good chance that within the next 10 to 20 years, we are going to see a massive change in what we consider meat and if it is even ethical to kill an animal to eat anymore. When we have an abundance of cheap plant-based meats, which is way healthier for you, gives you way more protein than you can get from animal meat then we are likely to move towards the idea that eating an animal is the same as cannibalism is to us nowadays. If you think back to 100, 200 years ago, there are a lot of cultures around the world that would not think twice about eating another human. Nowadays, it is pretty much unthinkable. I think we are going to see that shift in consciousness of humanity over the next 20-50 years towards eating animal proteins.

There is already a big veganism/vegetarian movement. I think we are going to see that completely take off as we move into the technology of plant-based meats. If we have vertical farms and plant-based meats, we do not need to raise sheep or cows. Maybe for milk, but we will see what happens in that space. It still means that much of the farmland is going to become redundant. If farmland becomes redundant, then it is not going to be worth the millions of dollars that it is now. It is going to become literally dirt cheap. At the moment, farmland is away from the cities. It is where people do not want to live. If we can open up these large tracts of land, and then 3D print a house really cheaply, and still have easy commuting distances because of flying cars and Hyperloop, then we are going to see the price of land drop dramatically. It is still early days yet to see exactly how this will play out.

This is one future that I am predicting, saying that farms will become redundant, and we will start to see a lot of farmland open up. Which, when they say: "We are not making any more land." No, we are not making any more land, but we are going to be repurposing it. It will become very cheap to do so because it is in the middle of nowhere. Because it is cheap, and there is no other purpose for it. There is no value added to it at all.

CASE STUDY

The Venus Project
www.thevenusproject.com

One interesting concept that is currently being experimented with is The Venus Project. They claim "The Venus Project proposes an alternative vision of what the future can be if we apply what we already know in order to achieve a sustainable new world civilisation. It calls for a straightforward redesign of our culture in which the age-old inadequacies of war, poverty, hunger, debt and unnecessary human suffering are viewed not only as avoidable, but as totally unacceptable."

The concept was thought up by Jacque Fresco and revolves around the idea of a resource based economy – one where money doesn't exist. Instead, it works with technology and automation to ensure that everyone can live a self-sustainable and abundant lifestyle. In fact, the founders believe in this future so much they have created an experimental city in Venus, Florida where it will feature intelligent houses; high-efficiency, non-polluting transportation systems; advanced computer technology; and a number of other innovations that can add value to the lives of all people.

What is really cool about the project is the way that it envisions its circular cities. In a well thought out plan, cities would be designed from the ground up and would be surrounded by parks and gardens. The central dome will house the core operations for the cities as well as educational facilities, health and childcare facilities. The next band is the cultural and entertainment centre, followed by the residential centres. Because of the circular nature of the city, streets would be easy to understand and you would never have to walk far to reach automated transport into the centre.

On the outskirts of the city will be hydroponic, aeroponic and aquaponic facilities and outdoor agricultural belts which will be used to grow a wide variety of organic plants. The outermost perimeter is utilised for recreational activities such as biking, golfing, hiking, and riding.

We are yet to see how this experiment will turn out, but I love the idea and cannot wait to see the progress they make.

UBI

One idea that is starting to be experimented with around the world is this idea of a universal basic income (UBI). Universal basic income is kind of like a citizen's salary. It is a basic income that is paid to every single citizen of that country, of that city, of that nation, with absolutely no obligations, no questions asked, no need to spend it in any particular manner.

There are a lot of arguments for and against UBI that we will look at, but this is where I talk about where the demonetisation of housing does not necessarily have to come from the housing side itself. It is the ability to have the house provided free for you. And so, if we have some sort of universal basic income, it means that everyone can afford at least that basic care of housing and essentially, it is the government or whatever regulatory body we will have, looking after us in the sense that that money then provides the accommodation for us, it provides food for us, it provides water, it provides the ability to create our own energy. Whether it is a tiny house or a regular house; imagine if the government made a lump sum payout of $10,000 at the start of every year and you could put that money towards buying a house, paying off a mortgage, all sorts of things. We have already talked about different blockchain applications that could come into play, so that money could simply just go into buying more and more shares of your own home.

Universal basic income is a big debate point, because no one quite knows what direction it would go, but there are experiments already in profess all around the world. One of the biggest leaders of these experiments is an organisation called Give Directly – **www.givedirectly.org**.

Some of the places that are experimenting with UBI at the moment are Utrecht in the Netherlands, Finland, Kenya, Namibia's having a play, Uganda, India, there are even certain organisations in USA and Canada that have been experimenting as well. Some of those experiments started way back in the '70s.

None of these experiments are on a big scale, they are mostly towns or villages. Utrecht is probably one of the biggest experiments out there, but even then, that was not to all the citizens of Utrecht. What is interesting is that one of the biggest arguments against universal basic income is that it will make people lazy. It will make people addicted to drugs, they will have money to spend on cigarettes and drinking and it will ruin family dynamics. But so far, the result from the experiments, particularly in Kenya, have been very positive. The organisation just simply gives a lump sum payment via mobile payment systems and they just ask a few questions, that's all. They do not dictate what the money is going to be spent on, they do not step in, they do not have any influence at all other than asking a few questions and what they have found is that, so far, it seems to be working.

UBI is predicted to bring with it economic growth, because a lot of people will start investing in themselves if they are not worried about surviving anymore. They are going to start thinking about how they can start excelling at what they are doing and going out and getting better jobs, investing in themselves to get a better life, to get a better degree, to become an entrepreneur, to start up their own business. Obviously, business then creates jobs and so there's this massive ability for the universal basic income to do really well.

While it is probably still a fair way off that houses will be completely free, there is definitely a path being laid that is looking like we are going to head in that direction, particularly for some sectors. Whether it is because of universal basic income, whether it is because energy becomes cheaper, whether it is because of the six D's of Digitisation, and then Deceptive, moving into all the way to the Demonetisation of housing, it definitely seems that we are on that pathway. The question then becomes how do you make money out of property as a real estate investor? What happens when you can get a brand new house for cheaper than a secondhand house, what happens when you can build a tiny house for one-tenth of the price of an existing house? What happens when your home is completely self-sufficient, you do not need to be connected to the grid?

What happens to councils when you no longer need to apply for building permits? What happens to all of this if farms become irrelevant and large tracts of land open up, that people can just simply move onto and start building on, start moving their tiny houses onto, and essentially start living free, what happens then? How do you make money? This is a question you need to start thinking about, because this trend is going to start affecting us in a variety of ways in the next 10 to 30 years. Immediately, some answers I can come up with are, are there things that you can do to make your house more self-sufficient now that a tenant would be willing to pay more rent for because they know that they are not going to have all these ongoing costs of living there?

Or, if you are buying and selling houses, is it something that you could get a higher price for, because the buyer knows that they are not going to have the ongoing costs, so they can afford more in mortgage repayments? What is it that you could be doing now to future-proof yourself in case house prices were to decrease? If house prices were to reduce to zero, because houses are not worth anything anymore, what is it that is worth something? This is where we go back to looking at things like community. Maybe someone does not want to pay to live in a house, because they can go out and create and live in their own house, but maybe they are willing to live in a community, or as in seasteading, they are willing to invest or pay to be part of a new country, a new city, where they are really starting to live on purpose rather than simply just living to survive.

Very big questions. Like I said, this is not going to have so much of a role until that 10 to 20-year mark, but there are definitely trends that we are starting to see and that are very much becoming cemented.

CHAPTER SEVEN

THE 3T SYSTEM FOR EXPONENTIAL SUCCESS

This can all be very mind-blowing because there is so much information here and it is kind of like, "Okay, which parts do I take notice of and which parts do I forget about? Which parts are going to influence me the most? Which parts can I take advantage of?" It's just getting crazy. So, if you think back to the very beginning, I told you there are two ways that you can take advantage of this information. You can either figure out what is going to influence your industry or your business the most, and then adapt to that particular trend or technology, or, you can find a new trend or technology that you really like and go jump into that, starting a new business or pivoting your existing business to be in line with that particular trend or technology.

So, take a minute to think to yourself, "What is it that I want to be doing? Do I want to be taking it and implementing it into an existing business or do I want to be changing my existing business towards the new trend or technology?"

The second thing you need to be doing is working within the scope of the three-T system. This is team, trends and technology.

I uncovered the 3T system after working with many mentors over the years and spending tens of thousands of dollars on business programmes. Through working with these top entrepreneurs, futurists and business people, I discovered that the part that most people tripped on was from going from being self-employed to running a team of people, with many entrepreneurs either giving up or continuing on really badly and never perfecting their business model. However, having a team by itself isn't enough, the entrepreneurs who really achieved fast growth and success were those who were leveraging the latest trends with new technologies. Therefore each T in the 3T system is just as important as the other:

Team: by having the right people, managing their teams, running their own projects. ***Trends:*** what are trends that are coming through and how can you leverage that using technology and using your teams?

Technology: what are the technologies that are emerging that are going to help you with what you're doing, more efficiently and more effectively?

Team

Gone are the days of building a top-down organisation. Your hierarchy needs to be horizontal. This is where you are starting up smaller globes of teams within your existing organisation and they are self-sufficient, making money and contributing to the greater organisation. In fact, Amazon's Jeff Bezos says: '"If you cannot feed your team with only two pizzas, you have got too many people in your team".

In the book 'From Good To Great' by Jim Collins, they say great companies started off by just bringing on the very best people, the people they knew they wanted to work with, and then steering the direction of the company from there. Can you do something similar to that? Bringing on the people you want to be working with and then seeing what direction you should be taking things? Is that going after a specific trend or after a specific technology?

You need to be able to give the team their own ability to build their business within your business. If you think about a company such as Amazon, Amazon succeeds because they have thousands of business owners leveraging their platform. These business owners want Amazon to be highly successful because it means they are going to be highly successful. Your remuneration package cannot be just a simple salary because that does not drive people anymore. It needs to have some sort of bonus or commission, letting people take charge of their own project within your company and then running with that, building that, ultimately turning it into its own company. This is your talent pathway. Don't bring someone in simply to do a job as they will leave the minute they can get a higher paying job or a job that provides them with more stimulation. Instead, create a pathway for how good team members can become intrapreneurs and build their own business within your business thereby contributing to the entire ecosystem.

Again, looking at Amazon, it has hundreds of departments within its company each running its own product, building its own business while staying aligned with the greater picture. From Amazon Web Services to Amazon Distribution, every single part of Amazon is run by someone else and every subpart of that company is then run by another team leader. That is the reason why it is now the most valuable company and it has made Jeff Bezos the richest man in the world. Because every single part has a team that is profitable. You do not simply have the marketing department and the sales department anymore. Every team has its own profitability where it is running and it is creating their own profit to run its own project.

You need to bring on team leaders who know how to run a project. They know how to run a promotion to create the funds that are needed not only for their own profit, but also to continue driving any innovation or any development within your company.

Trends

We have already spoken a lot about trends in this book and I really want to enforce and emphasise how important it is that you are not a follower of trends. You need to be one of the early adopters. When you see a trend coming up and you know it can be implemented well into your business, jump into it. Now, obviously, you do not want your whole businesses stake riding on one particular trend, so here is how you do it.

You set up a small task force within your existing business to go out and experiment with that trend. You do it in a way that, if that experiment fails, it is not going to impact the rest of your business. But, if that experiment works really well, it is something that you can adopt very quickly across the rest of your business. This is why it is important to be constantly experimenting within different parts of the business so that you can learn faster what works and what does not and move it across the whole organisation.

This ties in with team, where you have now got multiple teams doing multiple experiments within multiple parts of the business and you can leverage off of them. Leverage off the trends and act, this allows your organisation to gain a consciousness and a learning culture, with every part of the business sharing with every other part of the business. So, the business as a whole grows exponentially and it means you can sit there, as the founder, as the CEO, whatever position you take up overseeing it all. But, it means you are not the one having to do all the work, you are not taking all the risks because it's mitigated amongst every single little team taking on their own little experiments within the trends.

Technologies

Again, we have looked at many different technologies, so which is the technology that is right for you? Go out, see who the biggest players in that field are. Who are the guys innovating with technology? Who are the guys doing the best? Can you get to know them or find them on Facebook, LinkedIn, or Twitter? Connect with them. Send them a tweet. "Hey what you guys are doing is really cool," because if you can connect in with them, you might come across a partnership opportunity. There might be some way that you can get early access to their technology, which again allows you to be an immediate market leader. Now, in doing this with technology, you need to take the same approach as I was talking about within the trends section. You need to have just a small experiment with that technology to see if it is going to fail or to see if it's going to work. If it fails, burn it. It does not make a difference, or tweak it and see how you can make it work. If it does work, how quickly can you spread it across your entire organisation?

By following the three-T system, you are really going to be able to leverage team, technology, and trends to have a truly exponential company. This is pretty uncommon within the real estate sector, and there's only a handful of companies that have been able to leverage team, trends, and technology to become a global dominator within their space. Companies such as Airbnb, companies such as Matterport or Apis Cor.

You see, if you can remove yourself as the bottleneck within your business, you can truly grow exponentially. The old way of doing business is over. It is dead. The old way of thinking is so heavily ingrained within the real estate industry that it is stopping a lot of people from thinking exponentially. This is why I talk about the redistribution of wealth within the real estate industry that will start to happen over the coming decades, taking it from those who cannot pivot, those who cannot move fast enough, to those who have an economy of speed. This is again where we're talking about having the different teams within your organisation being able to run their own projects, taking advantage of many of these different aspects. These teams should also be working to disrupt each other, because if you are not disrupting yourself, then someone else will.

Level Up

We have covered a lot of very foreign concepts in this book that the average property entrepreneur probably has never thought about, or maybe has not even heard of before. In fact, your head might be spinning, wondering what to make of all this newly acquired information.

Most importantly, however, you are probably wondering how to apply many of the principles covered in this book in your business. We do this predominantly by focussing on the 3T system, which will take you from working a job in your property business into running a fully-fledged business utilising team, trends and technology, to create an organisation that can think and run by itself without you having to do all the heavy lifting.

See, I was the same when I started in business. I wanted to level up, so I set up my own business. It was still property-related and allowed me to keep in the game and know what was going on. I wanted to try and leverage my time and skills, so that I could keep investing in property but in a way that did not take up all my time and energy.

I tried bringing on apprentices and teaching them everything I knew with the aim of helping them to learn so they could help me to do more deals, this cost me a lot of time and did not result in any tangible outcomes.

All of this changed when I discovered exponential business models, and took on mentors who could show me how to leverage technology, trends and teams and scale it from a Minimum Viable Product that was already making money into a fully-fledged exponential business. I learned that if I leverage team, trends and technology then I can build a scaled business that does not rely solely on me.

I think the biggest kick up the ass I had was realising there are incoming waves that can be jumped on right now which will make 10 times more money with 10 times less effort. The longer

I waited before implementing these principles into my business, the more time I would waste and I would lose a tonne of money due to opportunity costs, plus my own assets would not be performing as well as they could have been. So, I had to act.

The interesting thing is some of the biggest 'A-ha!' moments I had was when I looked at the trends and waves and realised that anybody could do this.

You do not need to be tech-orientated - you are looking to utilise the technology not create it yourself. And, it makes no difference whether you want a large company or a small company, leveraging technology, team and trends still gives you the best return on your time investment. Itis like setting up the plumbing, but you can decide how much to turn the tap on.

By leveraging team, tech and trends there is no need for massive capital outlay.

But while anyone can do it, it needs to be done properly. I have seen too many property entrepreneurs out there making big mistakes in their business, simply because they follow what everyone else is doing, instead of being the one everyone else follows. Many of them try to do everything themselves and as a result get nowhere. Often, this leads to them being a step behind and trying to implement strategies that worked in the past, but do not work anymore. They do not even think about looking into the future to see what will be coming. The worst mistake of all I see many people making is playing not to lose, as opposed to playing to win.

These are such big problems because it all leaves you unaware of what to expect in your business and navigating the waters overcoming issues as they arrive, rather than looking forward, anticipating what issues are about to arise and instead of overcoming that issue, having a business that turns it into an opportunity and dominates because of it.

Most entrepreneurs are flying by the seat of their pants, living day by day and working IN the business, not ON the business. Therefore they get too caught up and cannot see the forest for the trees.

The biggest problem is that in order to level up, you need to unlearn and do everything differently than what worked for you to get to where you are now. It seems unnatural, but you need to give something up. Until you can do that, you will be stuck forever rowing the boat instead of captaining a ship that other people are sailing for you. You have probably tried hiring staff, or implementing certain technologies into your business and struggled.

I discovered early on in my entrepreneurial career the power of mentorship and to this day I continue to ensure I have someone mentoring me through business at all times. What I discovered is the magic that comes out of having someone else looking at my business, but also the power of being an active member of a highly engaged community. This is why I put together the Future 5 Mastermind to help entrepreneurs like yourself to leverage these 5 incoming waves of disruption, to take a different look at your property business and find the hidden opportunities in your portfolio and your industry, giving you the unfair advantage over your competition and catapulting you to the forefront of your industry.

You could go out and do it on your own but you would need to figure it all out by yourself, spend tens of thousands of dollars on mentoring and the amount of time it would take you to achieve the results would add up to hundreds of thousands of dollars in missed opportunity costs.

Or, you could surround yourself with people who are out there doing it, who are taking advantage of these trends and technologies and then sharing how they have done it and achieved those results.

They say that you are the sum of the five people you spend most of your time with. We are helping you to raise the calibre of those five people who are building exponential property businesses, by being part of a community of future facing property entrepreneurs, who are focussed on leveling up their businesses utilising the 3T system and growing together.

You will save years, if not decades, by connecting in with the smartest property entrepreneurs around and co-learning what it takes to have a successful property business in the 21st century.

It will give you the skills, knowledge and connections you require to level up your business, and command a team running an exponential property business that does not rely solely on you, nor require tonnes of time, capital or expenditure.

So, if you would like to know more about how you can join a group of property entrepreneurs, who are kicking ass applying the 3T system to their current businesses, whilst having me as your copilot in your business, checking in with you, helping to keep your business current and making the most money and utilising the latest trends possible, then head to **www.futureofpropertymastermind.com** to find out more.

Final Actions

In order to make the most out of this book and to continue learning, here is what you need to do:

1. Join the Future of Property Facebook group so that you can stay in the loop with many of these trends and technologies where I share regular examples of companies that are leveraging them and getting amazing results. Here you will be able to connect with other future-focussed property entrepreneurs.
   www.facebook.com/groups/1634080363532994

2. Follow the Freedom Investor Facebook page so you can stay up to date with all my new content on how you can apply the 3T system within your business and leverage the emerging trends and technologies. **www.facebook.com/thefreedominvestor**

If you are interested in learning more about my businesses and how you can become more involved:

1. The Freedom Investment Syndicate is part of my drive to ensure that everyone on earth can earn life-changing returns on their money which will help them to achieve financial security faster and more securely. We are still very much in the early stages, but we are getting our investors some amazing results. The minimum investment is $10,000, and if you would like to know more about the Freedom Investment Syndicate then email me directly at **kyron@futureofproperty.co** Subject: Freedom Investment Syndicate
2. The Global Lifestyle Club is an opportunity for entrepreneurs to own luxury properties around the world. We already have some pretty cool properties in our portfolio and would love for you to be a co-owner with us. If you have ever dreamed about owning a villa in Bali, a riad in Morocco or a farmer's cottage in Tuscany, then you definitely need to check out **www.globallifestyleclub.com** to download our information packs.

Thank you for taking the time to read this book, and more importantly thank you for being someone who is pushing the limits and making the world more awesome. There is nothing that excites me more than talking about world-changing ideas with other entrepreneurs and changemakers. The world is our oyster and anything is possible. I would love for you to join me on leaving a lasting impact on the world and fulfilling your legacy! If you loved this book then please reach out to me on social media or email. I love hearing from people like yourself.

To close out this book, I would like to leave you with my three favourite quotes that inspire me to wake up each morning and do awesome.

"Never doubt that a small group of thoughtful,
committed citizens can change the world; indeed,
it's the only thing that ever has"
—Margaret Mead

"Don't ask what the world needs.
Ask what makes you come alive, and go do it.
Because what the world needs
is more people who have come alive."
—Howard Thurman

"Those who are crazy enough to think they
can change the world usually do."
—Steve Jobs

Made in the USA
Monee, IL
04 November 2026

67387521R00075